for

Physics

Boca Raton, Florida

DISCLAIMER:

This QuickStudy® Booklet is an outline only, and as such, cannot include every aspect of this subject. Use it as a supplement for course work and textbooks. BarCharts, Inc., its writers, editors and designers are not responsible or liable for the use or misuse of the information contained in this booklet.

ISBN 13: 9781423202677
ISBN 10: 1423202678

Author: Mark D. Jackson, Ph.D.
Publisher:

BarCharts, Inc.
6000 Park of Commerce Boulevard, Suite D
Boca Raton, FL 33487
www.quickstudy.com

Printed in Thailand

Contents

Study Hints

NOTE TO STUDENT:

Use this QuickStudy® booklet to make the most of your studying time.

QuickStudy® samples offer easy-to-understand explanations and step-by-step instructions.

> **SAMPLE:**
> 0.085 => 0.08; 0.035 => 0.04 0.453 => 0.45
> 0.248 => 0.25

QuickStudy® tips alert you to common studying pitfalls; refer to them often to avoid problems.

> ⚠**PITFALL:** If the unit is wrong, the answer is wrong!

QuickStudy® notes provide need-to-know information; read them carefully to better understand key concepts.

> **NOTES:**
> - T(K) is always positive. Lab temperature is often measured in °C or °F, and must be converted to Kelvin for any calculations.
> - T always refers to Kelvin, unless specifically noted in the equation.

Take your learning to the next level with QuickStudy®!

1 Overview: What Is Physics All About?

- Physics seeks to understand the natural phenomena that occur in our universe.
- The goal of physics is to develop quantitative descriptions based on experimental measurement.

The motion of the planets in our solar system

Trajectory of a rocket or baseball

Rotational properties of the Earth

Flow of air and water through a pipe

Electrical circuits—voltage, current and resistance

A compass interacting with a magnetic field

Optical instruments such as telescopes and eyeglasses

Electrons in atoms and molecules

The wave nature of light

Understanding Physics

A description of a natural phenomenon uses many specific terms and definitions.

The mathematical equations define variables and constants as a means to predict and explain experimental behavior in the laboratory and our universe.

Solving Problems in Physics

A basic guide to solving problems is given in chapter 8, **Equations & Answers.**

In physics, we use the **SI units (International System)** for data and calculations:

Basic Physical Quantity	Symbol	Unit
Length	l, x	Meter - m
Mass	m, M	Kilogram - kg
Temperature	T	Kelvin - K
Time	t	Second - s
Electric Current	I	Ampere - A (C/s)

Other physical quantities are derived from these basic units. A list of many common **physics variables** is given in the appendix.

As a result of using the metric system in collecting data and performing calculations, we need to use prefixes for various units. **Unit prefixes** are summarized in the appendix.

Many variables and concepts are symbolized using Greek letters. The appendix gives the upper- and lowercase **Greek alphabet.**

Math Skills

Many physical concepts are only understood with the help of mathematical concepts from **algebra, statistics, trigonometry** and **calculus.**

Mechanics

Classical Mechanics

In **classical** or **Newtonian mechanics**, the position of a body is given by an equation of motion with position, velocity and acceleration as key variables.

The behavior is called **deterministic** because the future properties of the system are determined by current properties.

- **Mass** is the measure of the amount of matter.

  The standard unit for mass is kg
  1 kg = 1,000 g

- **Inertia** is an intrinsic property of matter; it occupies space.

- In solving a mechanics problem, the goal is to describe the future position, velocity and acceleration, given the properties of the body and the current coordinates, velocity (with direction) and acceleration.

Rectilinear Motion

Motion along a **straight line** is called **rectilinear** motion.

The equation of motion describes the position of the particle and its velocity in terms of the elapsed time, **t.**

The linear motion can be in one dimension (along a line), in two dimensions (on a plane) or in three dimensions.

Key Terms

- **Velocity (v)** is the rate of change of the displacement (s) with time (t):

$$v = \frac{ds}{dt} = \frac{\Delta s}{\Delta t}$$

- **Acceleration (a)** is the rate of change of the velocity with time:

$$a = \frac{dv}{dt} = \frac{\Delta v}{\Delta t}$$

NOTES:

a and **v** are vectors; it is important to know the magnitude ***and*** directional components.

The unit is also a key part of the velocity and acceleration:

Velocity –unit of meter/second (m/s)

Acceleration –unit of meter/second^2 (m/s^2)

- **Speed** is the absolute value of the velocity.

  - Speed equals the magnitude of the velocity vector; or the average distance traveled, divided by the elapsed time.

  - Speed is a scalar quantity with the **same units** as velocity, usually m/s.

$$\text{speed} = \frac{\Delta x}{\Delta t} \text{ (for 1 dimension)}$$

Motion in One Dimension (1-D)

Equations of motion are mathematical equations that describe the future position **(x)** and velocity **(v)** of a body in terms of the initial velocity **(v_i)**, position **(x_0)** and acceleration **(a)**.

- For constant acceleration, the position is related to the time and acceleration by the following equation of motion:

$$x(t) = x_0 + v_i t + \tfrac{1}{2} a t^2$$

- For constant acceleration, the velocity is related to the time by the following equation of motion:

$$v_f(t) = v_i + a t$$

- If the acceleration is a function of time, the position or velocity equation must be derived using the specific mathematical form for the acceleration: **a = a(t)**, instead of being a constant.

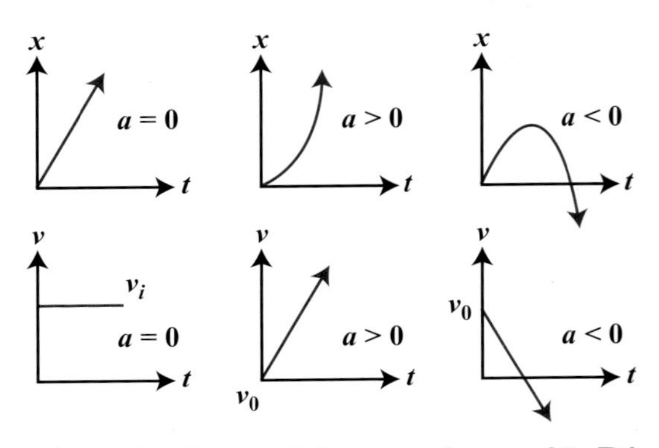

Motion in Two Dimensions (2-D)

- It is usually convenient to break the problem into separate equations of motion for the body moving on a plane (two dimensions).
- Select **Cartesian** or **polar coordinate,** depending on the type of motion.
- For bodies moving along a **straight line,** the best choice is usually a traditional **Cartesian coordinate** system, with variables ***x*** and ***y*.**
- This yields ***x*- and *y*-equations** of motion:

  $$\mathbf{x = v_{ix}\, t + {}^{1}/_{2}\, a_x\, t^2}$$

  $$\mathbf{y = v_{iy}\, t + {}^{1}/_{2}\, a_y\, t^2}$$

- Velocity and position are described using ***x*** and ***y*** components.

Cartesian

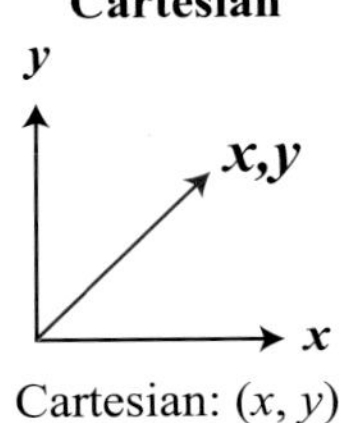

Cartesian: (x, y)

- For a **rotating body,** it is useful to use **polar coordinates,** with an angular variable, **θ**, and a radial distance from the rotational center.
- This yields equations of motion based on the angle and radial distance.
- The position and velocity have components that describe the rotational features of the motion.

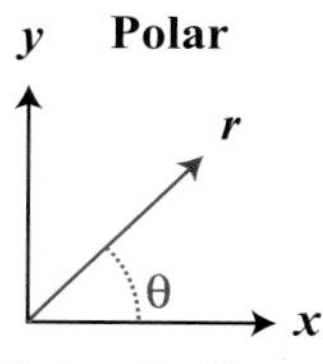

Polar: (r, θ)

Motion in Three Dimensions (3-D)

- Break the problem into separate equations of motion for each of the three dimensions.
- Select **Cartesian** or **spherical polar coordinates,** depending on the type of motion and nature of the forces.
- The physical interactions may dictate the best approach.
- For motion in a general Cartesian system, the ***x***- and ***y***- and ***z*-equations** of motion become:

$$\mathbf{x} = \mathbf{v}_{ix}\,\mathbf{t} + {}^{1}/_{2}\,\mathbf{a}_{x}\,\mathbf{t}^{2}$$
$$\mathbf{y} = \mathbf{v}_{iy}\,\mathbf{t} + {}^{1}/_{2}\,\mathbf{a}_{y}\,\mathbf{t}^{2}$$
$$\mathbf{z} = \mathbf{v}_{iz}\,\mathbf{t} + {}^{1}/_{2}\,\mathbf{a}_{z}\,\mathbf{t}^{2}$$

- Velocity and position are described using ***x*-, *y*- and *z*-components.**

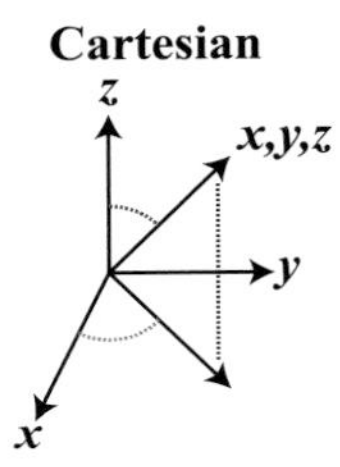

- For spherical coordinates, use equations of motion based on two angles, **θ** and **φ**, and a single distance, ***r***, the radial distance from the origin.
- The position and velocity have components that describe the **rotational,** or **spherical,** features of the motion.

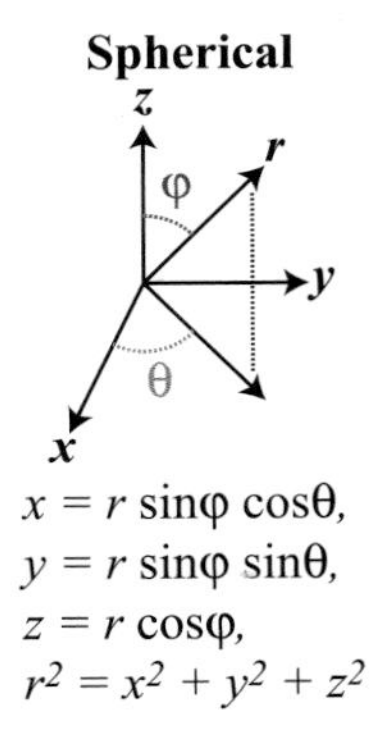

$$x = r \sin\varphi \cos\theta,$$
$$y = r \sin\varphi \sin\theta,$$
$$z = r \cos\varphi,$$
$$r^2 = x^2 + y^2 + z^2$$

Newton's Laws of Motion

Newton's Laws are the core principles for describing the motion of classical objects.

A body moves in response to a force. The Laws describe the response of a body to forces.

- The unit of force in the SI system is the **Newton, N,** defined as:

  $1\,N = 1\,kg\,m/s^2$

  A force of 1 N acting on a 1 kg body produces an acceleration of **$1 m/s^2$**.

- The unit in the cgs system is the **dyne:**

  $1\,dyne = 1\,g\,cm/s^2$

  A force of 1 dyne acting on a 1 g body produces an acceleration of $1 cm/s^2$.

- Forces can act on the **entire body** or the **surface** of the body. The former are described as forces acting-at-a-distance. Forces may also deform the shape or size of the body.

Newton's First Law

- A body remains at rest or in motion unless influenced by a force. This allows one to derive equations of motion for particles based on velocity, position and time.
- The **First Law** is one characteristic of **inertia,** a property of all matter.

Newton's Second Law

- The sum of forces acting on a body equals the mass multiplied by the acceleration.
- Force and acceleration determine the motion of a body, giving predictions for future position and velocity.
- The center of mass of the body is used to define the position of the body.
- The forces act on the center of mass of the body.

The mathematical form of the Second Law:

$\mathbf{F} = m\,\mathbf{a}$

or

$\Sigma \mathbf{F}_i = m\,\mathbf{a}$

NOTES:
Forces and acceleration are vectors! The net acceleration is determined by the **vector sum** of forces.

Newton's Third Law

- Every action on an object is countered by an opposing action.
- If two bodies interact, the force of body 1 on 2, $\mathbf{F}_{12}$, is equal to and opposite $\mathbf{F}_{21}$, the force exerted by body 2 on body 1.

$$\mathbf{F}_{21} = -\mathbf{F}_{12}$$

NOTES:
Notice the negative sign and the vector property of force. This means that forces always occur in pairs. For an object interacting with gravity, the downward force of gravity is balanced by a force exerted by the planet on the body.

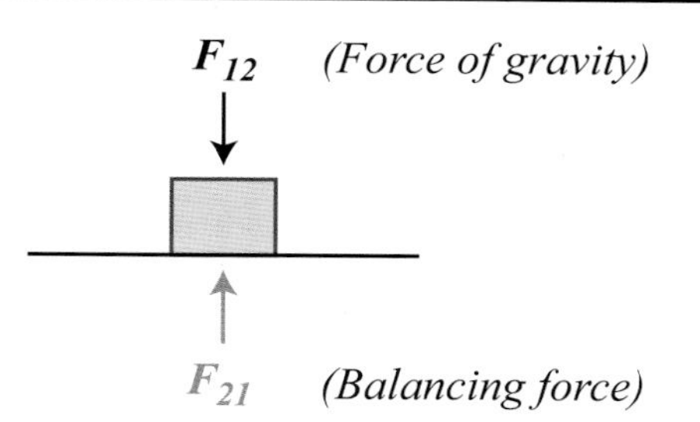

Types of Forces

A **body force** acts on the entire body, with the force acting at the center of mass of the body.

Consider the action of gravity on a body on the Earth's surface. The gravitational force, $\mathbf{F_g}$, pulls the object toward the center of the Earth:

$$\mathbf{F_g = mg}$$

The gravitational force is proportional to the mass of the body. The proportionality constant is **g:** the acceleration due to gravity, a constant at the Earth's surface. The constant **g** depends on the mass and radius of the Earth.

- **Weight** is a measure of gravitational force. This means that weight changes with altitude, or **g,** the gravitational acceleration constant.
- If there is minimal gravitational force, as in outer space, an object becomes "weightless."
- On the moon or other planets, weight also changes as **g** changes. The value of **g** on Mars or the moon would be less than **g** on the Earth's surface. As a result, the weight of an object would be different on the Earth, the moon and on Mars. The item would be lighter, reflecting the smaller value of **g.**

Table of g values for planetary bodies: unit of m/s^2	
Venus	8.87
Earth	9.8
Moon	1.62
Mars	3.71
Jupiter	23.1

- **Mass** is a measure of the quantity of material, independent of **g** (acceleration due to gravity) and other forces. The mass is measured in grams or kilograms.

Surface forces act only on the surface of the body.

- **Friction** is an example of a surface force. Friction, $\mathbf{F_f}$, is proportional to the force normal to the portion of the body in contact with a surface, $\mathbf{F_n}$.

$$F_f = \mu F_n$$

- **Frictional force** is usually proportional to the component of the gravitational force acting on the body, plus any additional force pushing down on the body.

  - **Static friction:** force resisting the movement of a body

  - **Dynamic friction:** force slowing down the motion of a body

For an object on a **horizontal** plane:

$$F_f = \mu F_n = \mu \, m \, g$$

Net force = $\mathbf{F_t}$ - $\mathbf{F_f}$

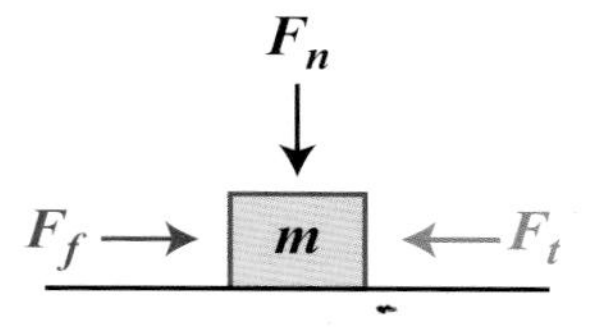

- The body moves only if the net force is positive (i.e., the applied force exceeds the resistant frictional force).
- For a body on an inclined plane, the normal force, $\mathbf{F_n}$, depends on the angle of the plane, with the normal, $\boldsymbol{\theta}$.
- The force of friction, $\mathbf{F_f}$, is always parallel to the plane and counters the component of the gravitational force, $\mathbf{F_t}$, that leads to the sliding motion of the body down the plane.

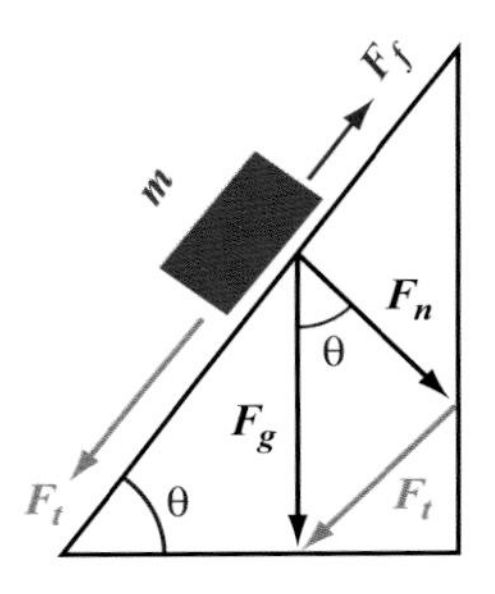

Circular Motion

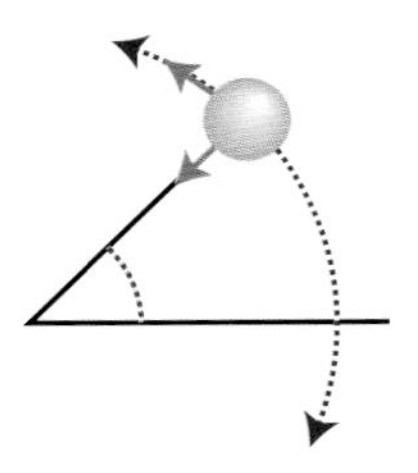

Goal: Examine a body moving in a circular path using two-dimensional **polar coordinates.** The motion is described using **(r, θ),** instead of **(x, y).**

Key Variables		
r	meter	the distance from the rotation center (center of mass)
θ	radian	the angle between r and the reference (x) axis
ω	radian/second	the angular velocity [Note: unit of rad/s, not cycles per second]
α	radian/second^2	the angular acceleration
s	meter	the circular motion arc **s = r θ** (**θ** in rad)

NOTES:
For a full rotation,
$s = 2\pi r = \pi \times$ diameter = the circumference of the circle

Tangential Acceleration & Velocity

$v_t = r\omega$; $a_t = r\alpha$

v and **a** along the path of the motion arc

Centripetal Acceleration

$$a_c = \frac{v^2}{r}$$

a is directed toward the rotational center

The centripetal force toward the center keeps the body in circular motion with a tangential acceleration and velocity.

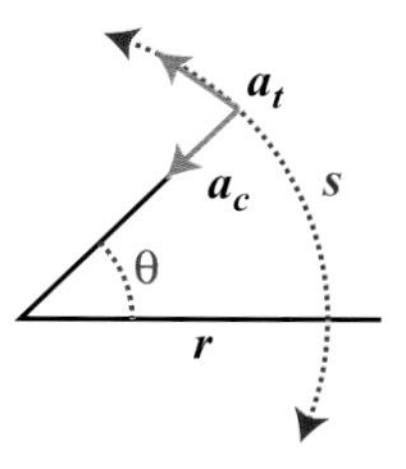

Kinetic Energy & Work

- **Kinetic energy** is the energy associated with motion of the body of mass **m** with constant velocity **v.**

  The symbol for kinetic energy is **K.**

  $$\mathbf{K = {}^{1}/_{2}\, mv^2}$$

NOTES:
The standard energy unit is the **Joule (J):**
  $\mathbf{1\ J = 1\ kg\ m^2/s^2}$
Mass is measured on kg, velocity in m/s, so K has a unit of Joules.

- **Momentum:** a property of motion, defined as the product of mass and velocity.

  The symbol for momentum is **p.**

  $$\mathbf{p = m\ v}$$

- **Velocity** is a **vectoral** property (with magnitude and direction); therefore, momentum is also described using a vector: **v** has a magnitude and directional components—$\mathbf{v_x}$ for one dimension, $\mathbf{v_x}$ and $\mathbf{v_y}$ for two dimensions, and $\mathbf{v_x}$, $\mathbf{v_y}$ and $\mathbf{v_z}$ for three dimensions.

The magnitude of the velocity:

$= |v_x|$ **for 1-D**

or

$= +\sqrt{(v_x^2 + v_y^2)}$ **for 2-D**

or

$= +\sqrt{(v_x^2 + v_y^2 + v_z^2)}$ **for 3-D**

- **Mass** is always a scalar quantity.

- **Work:** the force acting on a body moving a distance.
  For a general force, **F**, and a particle moving along a path s:

  $$\text{Work} = \int F\, ds$$

  Or, for motion in a Cartesian coordinate system:

  $$W = \int F_x\, d_x + \int F_y\, d_y + \int F_z\, d_z$$

  For a **constant force,** work is the scalar product of the two vectors: force, **F**, and path, **r**:

  $$W = F\, d \cos(\theta) = F \bullet r$$

NOTES:
F and **r** are vectors forming angle **θ**.

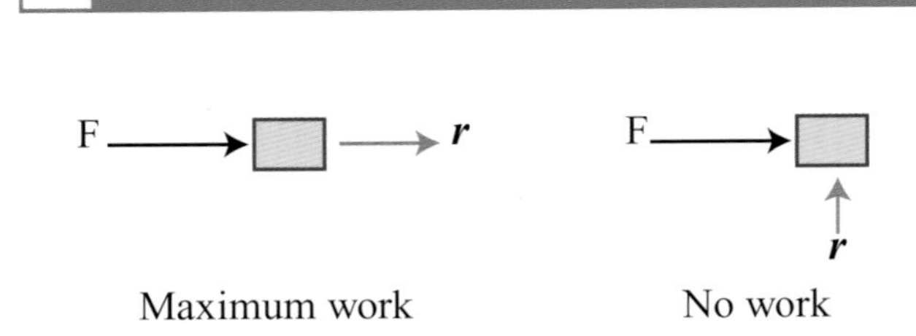

W is maximum for $\theta = 0$

In this case the force and motion are in the same **direction:**

$W = f \times r$, since $\cos(\theta = 0) = 1$

W is zero (minimized) for $\theta = 90$ degrees $= \pi/2$.
No work is done if the force is normal to the motion:

$W = f \times r \times 0 = 0$, since $\cos(\theta = 90 = \pi/2) = 0)$

- **Power** is the work or energy expended per unit time, and measures how quickly work or energy is used.

  The symbol for power is **P.**

$$P = \frac{\Delta \text{Work}}{\Delta \text{time}} = \frac{\Delta \text{Work}}{\Delta t}$$

$$\text{Work} = \int P(t)\, dt$$

- The unit for power in the SI system is the **Watt,** symbolized by **W.**
- 1 watt of power is the expenditure of 1 Joule of energy in 1 second.

1 watt = 1 Joule / second = 1 J/s

- Power may vary with time, so work will also depend on the properties of power.
- Work for a **constant output of power:**

$$\text{Work} = P\,\Delta t$$

- If power **varies with time:**

$$\text{Work} = \int P(t)\,dt$$

- In some processes, work comes from the dissipation of kinetic energy:

$$W_{net} = K_{final} - K_{initial}$$

- A portion of the kinetic energy is converted to work.

Potential Energy

The total energy of a body is the sum of kinetic, **K,** and potential energy, **U:**

$$E = K + U$$

Energy conservation principles are used to study the interplay of potential and kinetic energy.

- **Potential energy** arises from the interaction of the body with a potential from an external force. The potential depends on the position of the body relative to the source of the force. Potential energy will always depend on this distance.

- **Potential energy** is energy of position: **U(r).** The mathematical form of **U** depends on the type of force that generates the potential. Consider gravity and electrostatics as examples.

- **Gravitation** $\mathbf{U(h) = mgh}$
  Body of mass **m,** a distance **h** above the Earth's surface, interacts with the mass of the Earth.

- **Electrostatic** $U(r_{12}) = \frac{q_1 q_2}{r_{12}}$

  Charge q_1 interacts with charge q_2 at a distance r_{12}.

- The total energy is given by summing the kinetic and potential energy terms:

  $$E = K + \sum U$$

Conservation of Energy

- The energy may have more than one potential energy contribution. For example, a charge may interact with several other charges; a planet interacts with the sun, as well as other planets.
- If there are no other forces acting on the system, the **total energy is constant** and the system is called **conservative.**
- Consider the motion shown in the figure below. Examine the motion at various points and determine the **K** and **U:** Points 1, 2 and 3 will correspond to different **K** and **U,** but the **total energy** is constant.

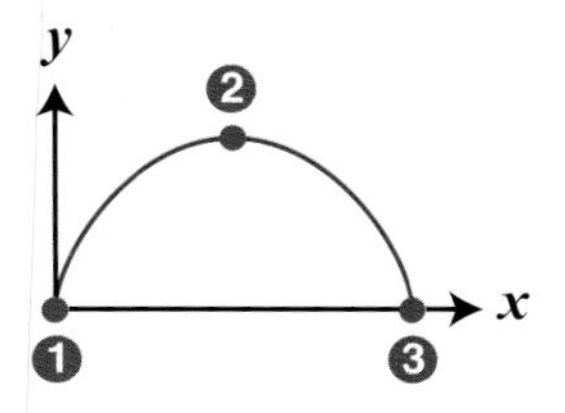

Collisions & Linear Momentum

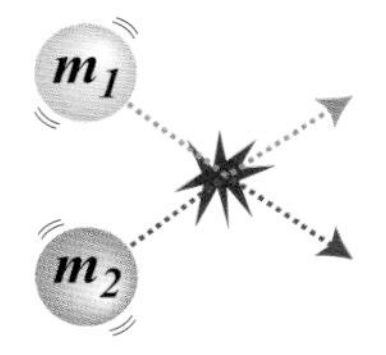

Goal: Examine momentum of colliding bodies.

Hint: For 2-D or 3-D, break the problem into Cartesian components. Examine the ***x*-, *y*-** or ***z*-** components of the forces: Each component must balance in order for the system to remain in equilibrium.

Key Variables & Equations

Types of Collisions:

- **elastic:** conserve energy
  Conservation of kinetic energy: $\mathbf{K}_{\text{initial}} = \mathbf{K}_{\text{final}}$
  Conservation of momentum: $\mathbf{p}_{\text{initial}} = \mathbf{p}_{\text{final}}$

- **inelastic:** energy is lost as heat or deformation

Relative Motion & Frames of Reference

A body moves with velocity **v** in frame **S;** in frame S' the velocity is **v';** if $\mathbf{V_s}$**'** is the velocity of frame **S'** relative to **S,** therefore, $\mathbf{v} = \mathbf{V_s}' + \mathbf{v}'$.

Key Equations

Linear Momentum: $p = m\,v$ (vectoral property)
Kinetic Energy: $K = \frac{1}{2} m\,v^2$ (scalar property)

- **Conservative Systems:** The elastic collision conserves kinetic energy, **K,** and momentum, **p,** since there are no external forces on the system.
- This means that there is no acceleration for the bodies.
- **Elastic Collision** – *conservation of kinetic energy*

$$\sum \tfrac{1}{2}\,m\,v_i^2 = \sum \tfrac{1}{2}\,m\,v_f^2$$

Examine *x-,y-* and *z-components* of the velocity.

- **Elastic Collision** – *conservation of momentum*

$$\sum m\,v_i = \sum m\,v_f$$

For 2-D (or 3-D), examine ***x-,y-*** and ***z***-components. Normally, collisions are confined to a plane.

- **Inelastic Collision** – *energy is dissipated in the collision*

For a **perfectly inelastic collision** in one dimension, bodies collide and stick together; *the momentum is conserved, but not the energy.*
$m_1 v_{1i} + m_2 v_{2i} = (m_1 + m_2) v_f$

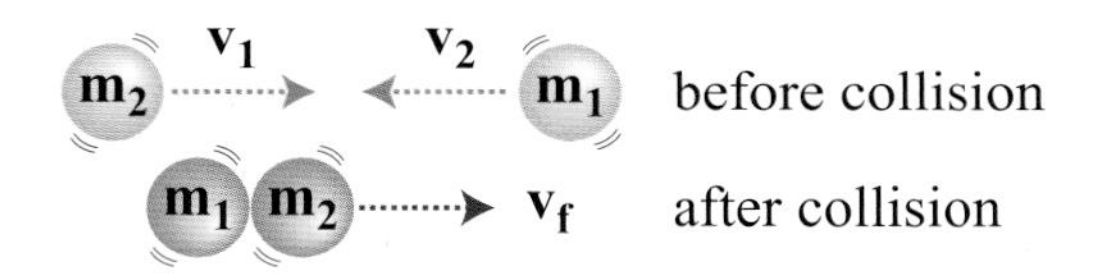

- **Impulse** is a force acting over time. The force can be constant or can vary with time.

Impulse = F Δt
or
$\int F(t)\,dt$

- Impulse is also the **momentum change:**

Impulse = p_{final} - $p_{initial}$

Rotation of a Rigid Body

Goal: Examine the rotation of a rigid body of a defined shape and mass.

Hint: This is constrained motion—the body only rotates about one axis; otherwise, the body is at a fixed position in space, the center of mass of the body.

Key Variables & Equations

- **Center of mass:** the "average" position in the body, accounting for the object's mass/position distribution.

  A body force, such as gravity, acts on the total mass of the object at this point. These coordinates are calculated from the mass distribution, the various mass elements weighted by position.

- **Center of mass:** x_{cm}, y_{cm}, z_{cm}

$$x_{cm} = \frac{\sum m_i x_i}{\sum m_i} \quad y_{cm} = \frac{\sum m_i y_i}{\sum m_i} \quad z_{cm} = \frac{\sum m_i z_i}{\sum m_i}$$

- For two balls connected by the rod, the center of mass depends on the mass of the balls. The center of mass is closer to the larger mass.

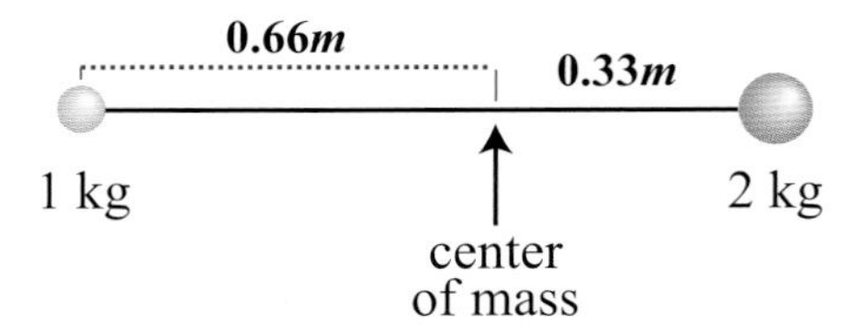

- The **Moment of Inertia** is a measure of the distribution of the mass about the rotational axis.

  Its symbol is **I.**

  I accounts for the fact that the rotational motion depends on ***how*** the mass is distributed around the axis, *not* the **total** mass of the body.

$$I = \sum m_i r_i^2$$

$\mathbf{r_i}$ is the radial distance from the mass element $\mathbf{m_i}$ to the rotational axis.

I must be defined for a **specific rotational axis** passing through the center of mass.

I accounts for the mass distribution relative to the axis of rotation.

I for bodies of mass **m:**

- **rotating** cylinder (radius R): $^1/_2$ **m R²**
- **twirling** thin rod (length L): $^1/_{12}$ **mL²**
- **rotating** sphere (radius R): $^2/_5$ **m r²**

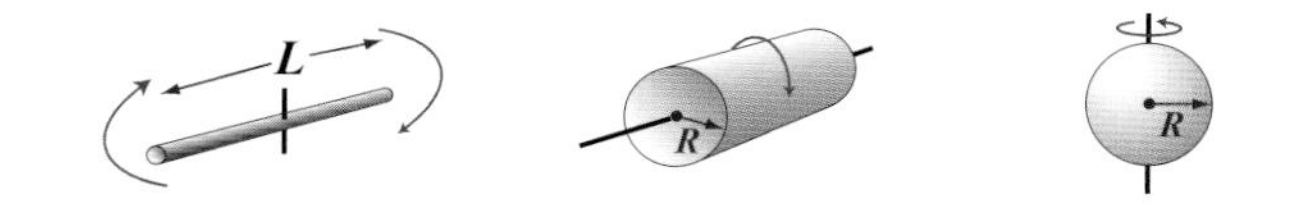

NOTES:
The rotational axis must be specified for the rod and cylinder. The sphere rotates about its center.

Goal: Quantify the force, energy and momentum of rigid rotating bodies.

■ Rotational Kinetic Energy $= \frac{1}{2} I\, \Omega^2$

The rotational energy varies with the rotational velocity and moment of inertia, **I.**

I functions as the **effective mass** for rotational kinetic energy and momentum.

- **Angular force** is defined as **torque,** and given the symbol **τ:**

$\boldsymbol{\tau} = \mathbf{I}\,\boldsymbol{\alpha} = \mathbf{r} \times \mathbf{F}$ (angular acceleration force)

- This is analogous to the **F** = **ma** for Newton's Second Law; in this case, the mass is the rotational mass, **I** (the moment of inertia), and **α** is the rotational acceleration.
- **Angular Momentum** is the momentum associated with rotational motion.
- Axis of rotation specifies the rotational motion, which passes through the center of mass of the body.

Angular momentum is given the symbol **L:**

$$\mathbf{L} = \mathbf{I}\,\omega = \mathbf{r} \times \mathbf{p} = \int \mathbf{r} \times \mathbf{v}\, d\mathbf{m}$$

- This is analogous to the **p** = **mv** definition for momentum; in this case, **I** is the rotational mass and **ω** is the rotational velocity.

NOTES:
L is a vector cross product, so angular momentum is also a vector.

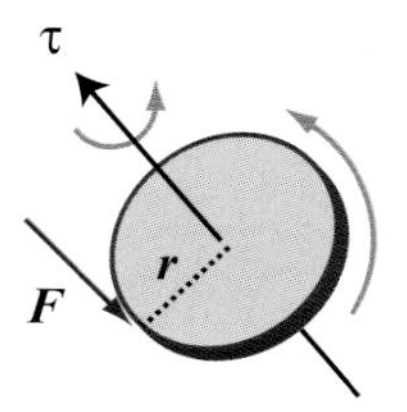

■ **Torque** is also the change in angular momentum with time.

$$\tau = \mathbf{r} \times \mathbf{F} = \frac{\mathbf{dL}}{\mathbf{dt}}$$

NOTES:
$\boldsymbol{\tau}$ is a vector cross product, so torque is also a vector.

Static Equilibrium & Elasticity

Goal #1: Examine several forces acting on a body.

Guiding Principles: Equilibrium is achieved when the sum of forces acting on a body is zero, and the sum of torques acting on a body is zero:

$$\sum \mathbf{f} = 0 \qquad \sum \tau = 0$$

The body has no linear or angular acceleration.

For the case of two bodies attached by a bar, equilibrium is achieved when $\mathbf{m_1\ x_1 = m_2\ x_2}$.

The point of balance is the center of mass.

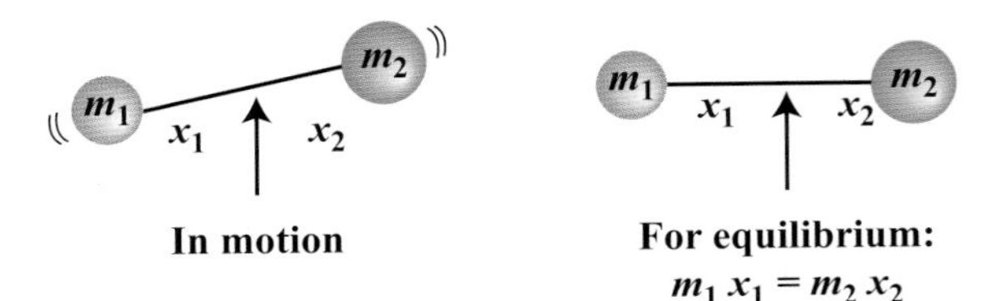

Hint: Evaluate each component; any net force moves the object; any net torque rotates the object.

Goal #2: Examine deformation of a solid body; elasticity of materials.

- **Elasticity** is the property of a material to return to its original shape after the force acting on it is removed.
- **Stress & Strain**

Strain, denoted by **ε**, is the deformation of the body, **Δ**, per unit length, **L:**

$$\varepsilon = \frac{\Delta}{L}$$

Stress is the force per unit area on the body that produces the deformation. Normal stress, denoted **σ,** acts perpendicular to the surface:

$$\sigma = \frac{dF_n}{dA}$$

Shear stress, denoted **τ,** acts tangentially to the surface:

$$\tau = \frac{dF_t}{dA}$$

- **Hooke's Law:** For elastic bodies, the stress is **linearly proportional** to the strain:

  Stress = constant × Strain

  $\sigma = E \times \varepsilon$

  E, the elastic modulus, is a constant that depends on the material.

Key Equation: Stress = elastic modulus x strain.

Elastic Modulus = stress/strain = force/change (Hooke's Law)

Linear Stress (Tensile): E = Young's Modulus, symbolized **Y:**

$$Y = \frac{\frac{F_l}{A}}{\frac{\Delta l}{l_0}}$$

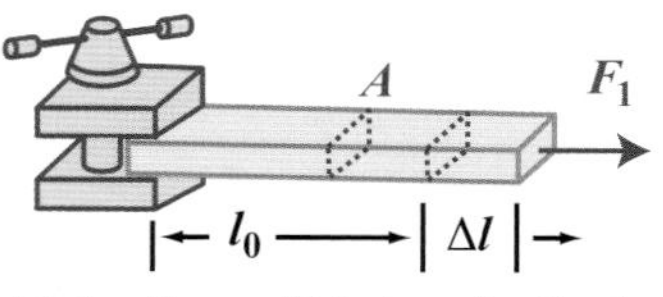

Note: Force F_1 is longitudinal

NOTES:
The force F_1 is longitudinal to the face A.
Think of a guitar string or piano string held in tension.

Shape Stress: E = Shear Modulus, symbolized S:

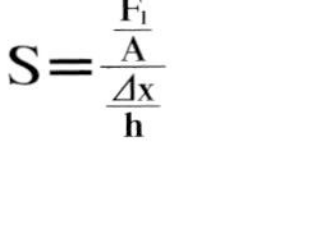

$$S = \frac{\frac{F_t}{A}}{\frac{\Delta x}{h}}$$

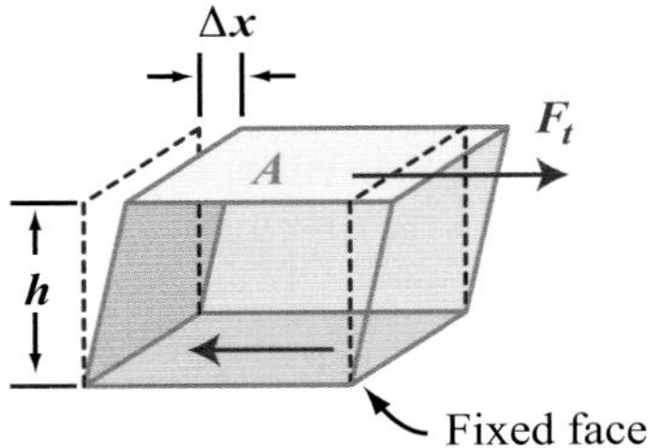

Note: Force F_t is tangential to face *A*

NOTES:
The force F_t is tangential to face A.
Think of wind shear, twisting motion of air in the atmosphere.

Volume Stress: E = Bulk Modulus, symbolized B:

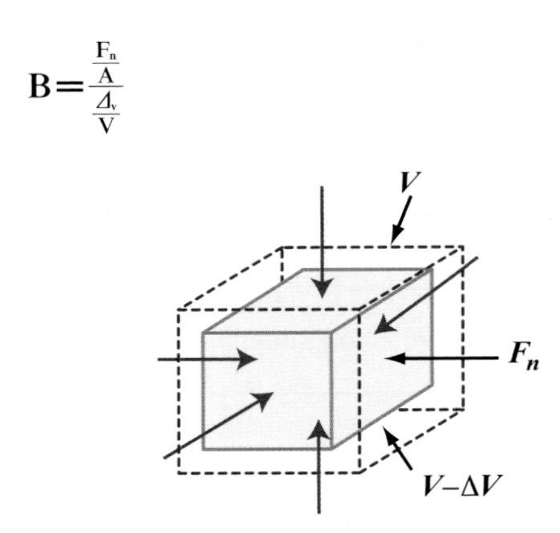

$$B = \frac{\frac{F_n}{A}}{\frac{\Delta_v}{V}}$$

Note: Force F_n is normal to face *A*

NOTES:
The force F_n is normal to face A.
Think of a uniform pressure compressing the body.

Universal Gravitation

Goal: Examine gravitational energy and force.

- All bodies interact—gravitational attraction is a property of mass/inertia.
- Planets are attracted to other planets, and objects on the surface of a planet are held in place by gravitational force generated by the planet.

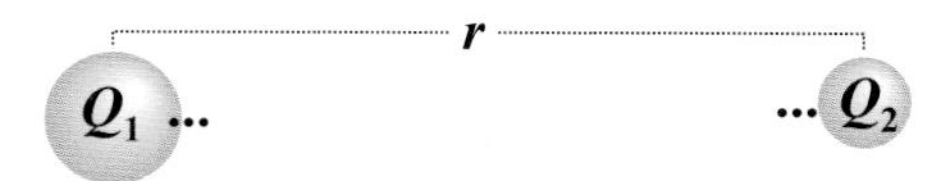

Case #1: Bodies of mass M_1 & M_2 separated by r interact via gravity.

Key Equations for Gravitational Force & Energy

Gravitational Energy: $U_g = \frac{GM_1 \times M_2}{r}$

For a group of bodies, $U_g = \Sigma U_{gi}$.

Each pair of bodies will interact and contribute to the gravitational energy.

Gravitational Force: $F_g = \frac{GM_1 M_2}{r^2}$

$\mathbf{F_g}$ is a vector, directed along r, connecting M_1 and M_2
For a group of bodies, $F_g = \sum f_g$

The net force is a vector sum of each f_g

Acceleration due to Gravity & Newton's Second Law

For an object on the Earth's surface, F_g can be viewed as $\mathbf{F_g = m\ g}$; **g** is the acceleration due to gravity on the Earth's surface.

Therefore, g is given by

$$g = \frac{GM(\text{earth})}{r(\text{Earth})^2}$$

For objects on the Earth's surface, **g = 9.8 m/s²:**
***g** will have **different values** on the moon, Mars and other planets.*

Case #2: A body interacts with the Earth, from a position relative to the Earth's surface.

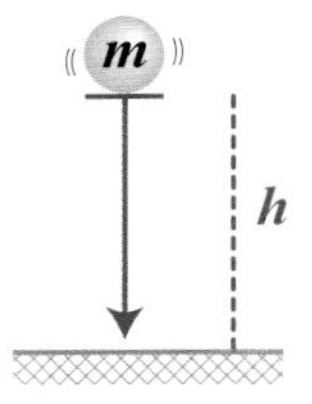

Key Equation for Gravitational Potential Energy: The mass of the Earth generates a gravitational potential. The zero of the potential is the Earth's surface.

Gravitational potential energy: $U_g = mgh$

For a body on the Earth's surface, $h = 0$; therefore $U_g = 0$

U_g can be calculated for any body of mass m at height h above the Earth's surface using $U_g = mgh$.

Weight is the gravitational force exerted on a body by the Earth.

Weight $= F_g = mg$

NOTES:
Remember: mass and weight are different. **Mass** is a property of matter. **Weight** is a measure of gravitational force on the body.

Oscillatory Motion

Goal: Study motion & energy of oscillating body.

Simple Harmonic Motion (1-D)

Consider a body of mass, **m,** attached by a spring to a non-moving wall.

- Force: $\mathbf{F = -k\Delta x}$ (Hooke's Law)
- Potential energy: $\mathbf{U_k = \frac{1}{2}k\Delta x^2}$
- Frequency of the oscillation: $\mathbf{f = \frac{1}{2\pi}\sqrt{\frac{k}{m}}}$

HOOKE'S LAW

Simple Pendulum

Consider a body of mass, **m,** attached by a rigid rod of length, **l,** free to swing back and forth:

- Period of oscillation: $\mathbf{T=2\pi\sqrt{\frac{l}{g}}}$
- Potential energy: U_g = mgh (h is the height above the lowest point)
- Frequency of the oscillation: $\mathbf{f=\frac{1}{2\pi}\sqrt{\frac{g}{l}}}$

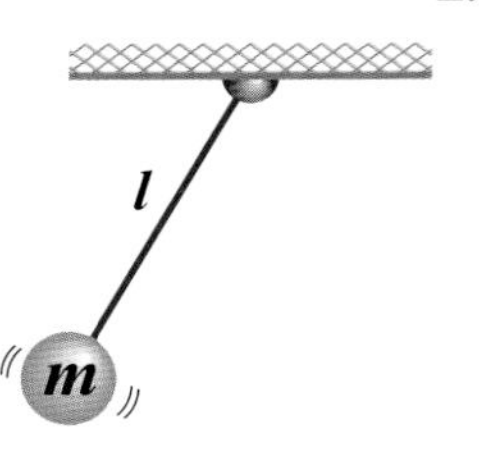

NOTES:
The frequency and period of the motion only depend on the length of the rod connecting the mass to the surface, and are independent of the mass of the object held by the rod. This is why pendulums are used in some clocks.

Forces in Solids & Fluids

- **A fluid—gas or liquid—takes the shape of the container.**

- **A gas fills the container.**

- **A solid has a defined size and shape.**

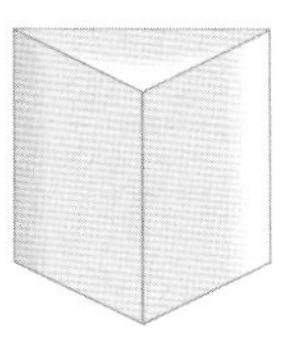

Goal #1: Examine properties of **static** solids & fluids.

- The **density** of a solid, gas or liquid is the mass of the material divided by the volume.

  ρ = mass/volume

- Mass is measured in g or kg.
- The volume is given in cm^3, L or m^3.
- Note the unit equivalence: 1 g/cm^3 = 1,000 kg/m^3.
- Common solids and liquids have densities ranging from 0.7–20 g/cm^3.

- Gas densities are roughly 1/1000th of solid and liquid densities.
- In addition, gas densities vary more with temperature and pressure than densities of liquids and solids.
- The **Ideal Gas Law** is an equation of state for gases; it predicts the relationships between **density, pressure** and **temperature.**

Table of Densities	Unit g/cm^3
Water	1.00
Benzene	0.88
Air (300 K, 1 bar)	1.16×10^{-3}
Iron	7.87
Lead	11.35
Magnesium	1.74
Gold	19.32
Aluminum	2.70
Platinum	21.45

Pressure in Fluids

Pressure is the force exerted by a fluid on the walls of the container. It is given the symbol **P** and defined as the force exerted divided by the area of the forces acts upon.

P = force/area

- The unit of pressure is the **Pascal,** abbreviated Pa.

$1\ Pa = 1\ N / m^2$

- Pressure x volume has units of energy, the **Joule.** So, in a sense pressure can be viewed as **energy density.**

Other Common Pressure Units & Conversion Factors	
Atmosphere	1 atm = 1.101325×10^5 Pa
Bar	1 bar = 1×10^{Pa}
mm Hg	1 atm = 760 mm Hg
lb/in^2	1 bar = 14.50 lb/in^2

- For gases, pressure arises from the motion of gas atoms/molecules colliding with the walls of a container.
- For liquids, pressure is due to the atmosphere pressing on the surface of the liquid and from the gravitational force exerted on a liquid.

- **Pascal's Law:** For a fluid enclosed in a vessel, the **pressure** is equal at all points in the vessel. This Law applies to liquids and gases.
- For a gas sample, it makes sense that the pressure is uniform throughout the balloon or container.
- This Law also finds application in dealing with hydraulic pumps, presses and other devices that used an enclosed fluid to transmit pressure.
- Consider the operation of a hydraulic press—liquid sealed in a container with two movable pistons: The pressure is constant at all points. However, the force depends on the area the pressure is exerted upon. **P = A F** for the enclosed liquid, where A is the area of a movable piston in the pump. If the two pistons have different areas, pressure on one piston translates to the same pressure on the other piston. But the force will depend on the area of each piston. The lifting capacity and, therefore, the work that can be performed with the press depends on the areas of the two pistons.

Relative Force Equation: $A_1 \ F_1 = A_2F_2$;

- The relative cylinder areas determine the relative forces on each piston.
- A small force on a small piston is translated into a larger force on a larger piston. But the pressure is the same throughout the pump.

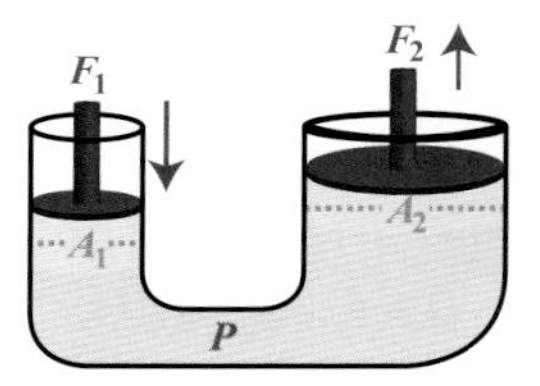

- **Pressure Variation with Depth:** A column of water generates pressure, in addition to the pressure of the atmosphere pressing on the surface of the liquid. This means that in a body of water, such as a lake or the ocean, the pressure increases with depth.
- The pressure below the surface, at depth h, has two contributions, the atmospheric pressure at the surface, P_1, and the weight of the liquid column, $\rho g h$.
- **Pressure Beneath the Surface of the Liquid**

$$P_2 = P_1 + \rho g h$$

h is the distance, or depth, beneath the surface of the water.

ρ is the density of the water, and **g** is the acceleration due to gravity.

In this case, we are describing the gravitational force acting on the entire column of liquid, not just the atmospheric pressure acting on the surface of the liquid.

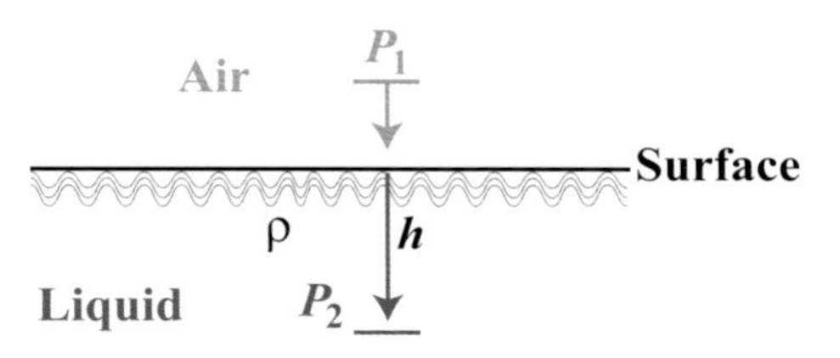

- **Archimedes' Principle:** An object submerged or partially submerged in water will feel a force, called the **buoyant force,** that tends to force the object to the surface and out of the water.

■ Buoyant force, F_b, acting on the object of volume V submerged in liquid of density **ρ** is given by the following equation:

$$\mathbf{F_b = \rho V g}$$

g is the acceleration due to gravity

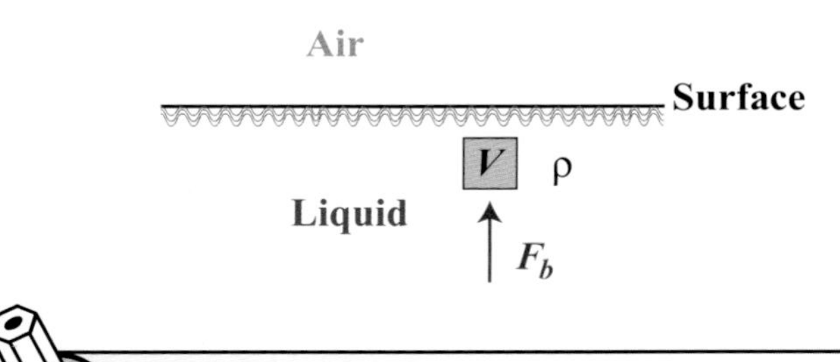

NOTES:
ρ is the density of the liquid, not the density of the object. **V** is the volume of the object and, therefore, the volume of the liquid displaced by immersing the object. The buoyant force matches the gravitational force acting on the water displaced by the object.

Goal #2: Examine fluid **motion** & fluid **dynamics.**

■ Properties of an **Ideal Fluid:**
- ◆ Nonviscous: minimal interaction between particles.
- ◆ Incompressible: the density is constant.
- ◆ Steady flow.
- ◆ No turbulence.

■ At any point in the flow, the product of area and velocity is constant:

$$\mathbf{A_1 \, v_1 = A_2 \, v_2}$$

This relationship is derived from the fact that fluid flow must conserve mass. Mass does not change if you move the fluid.

- **Variable fluid density:** If the density changes, the following equation describes properties of the fluid:

$$\rho_1 A_1 v_1 = \rho_2 A_2 v_2$$

- A sample application is gas flow through a smokestack; mass is still conserved, though density may change.

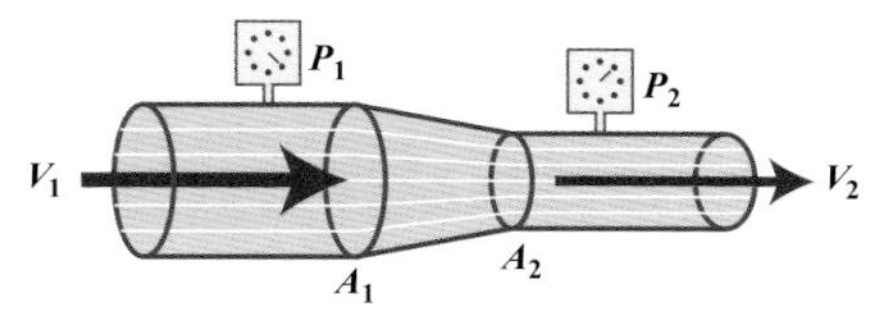

Flow Through a Hose

- **Bernoulli's Equation** is a more general description of fluid flow.

For any point y in the fluid flow,

$$P + \tfrac{1}{2}\rho v^2 + \rho g y = \text{constant}$$

This equation accounts for the conservation of energy associated with flow: kinetic energy and gravitation potential energy.

Special case: For a fluid at rest:

$$P_1 - P_2 = \rho g h$$

3 Wave Motion

Wave Properties

■ **A number of types of waves are encountered in physics:**

- Transverse
- Longitudinal
- Traveling
- Standing
- Harmonic
- Quantum mechanical

■ General form for **transverse traveling wave:**
$y = f(x - vt)$ (to the right)
or
$y = f(x + vt)$ (to the left)

■ General form of **harmonic wave:**
$y = A \sin(kx - \omega t)$ or $y = A \cos(kx - \omega t)$

sin and cos waves

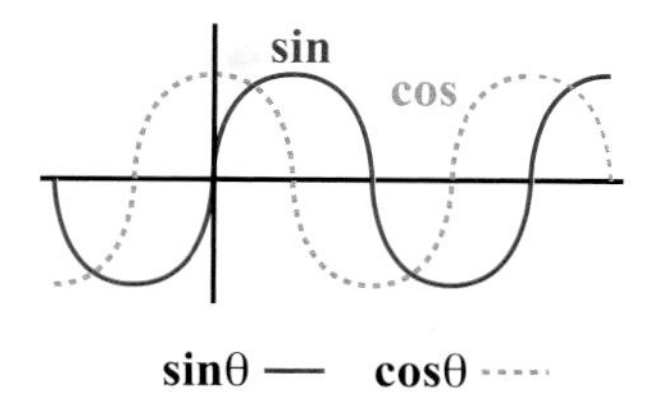

sinθ — cosθ ·····

■ **Standing wave:** Integral multiples of **λ/2** fit the length of the oscillating material. The wave is trapped in the box or along the wire.

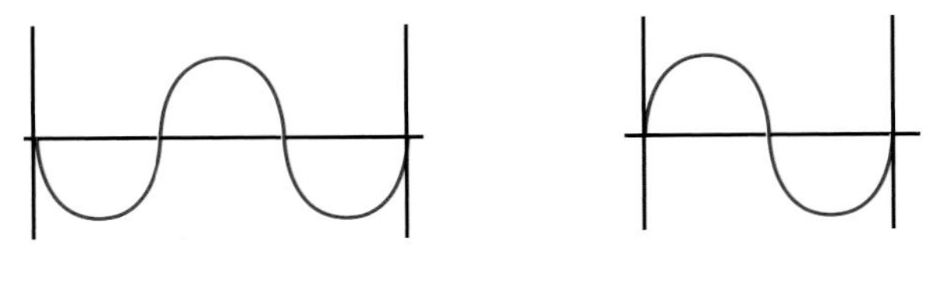

3/2 λ 1 λ

■ **General Wave Equation:**

$$\frac{d^2y}{dx^2} = \frac{1}{v^2}\frac{d^2y}{d^2t}$$

■ **Superposition Principle:** Overlapping waves interact, yielding constructive and destructive interference.

■ **Constructive Interference:** The wave amplitudes add up to produce a wave with a larger amplitude than either of the two waves.

Consider waves y_1 and y_2: The addition of these two waves gives the wave y: $y = y_1 + y_2$

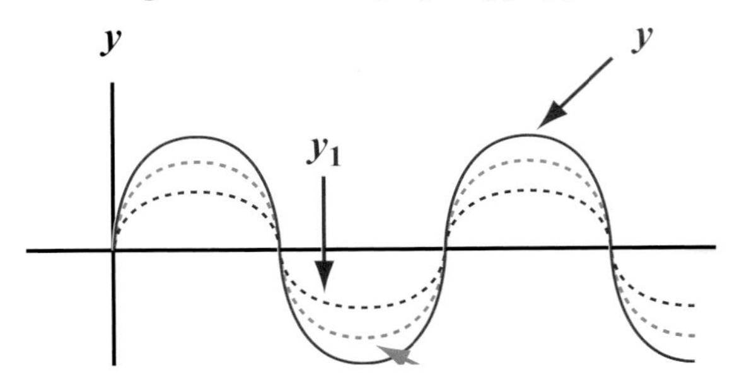

- **Destructive Interference:** The wave amplitudes add up to produce a wave with a smaller amplitude than either of the two waves. In this case, the two waves cancel out.

Consider waves y_1 and y_2: The addition of these two waves gives the wave y: $y = y_1 + y_2$

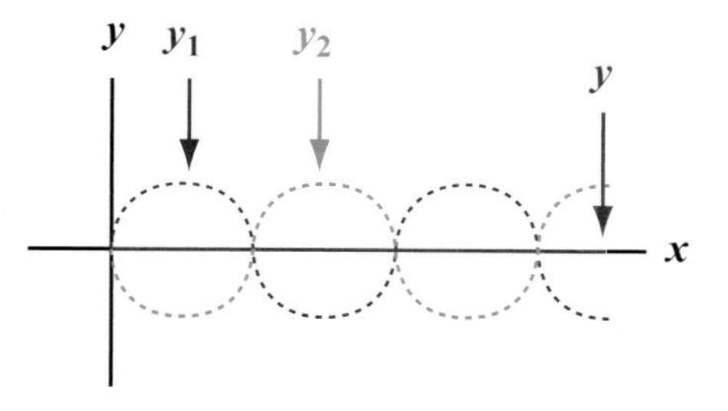

- **Harmonic Wave Properties:** Harmonic waves have a number of parameters that allow you to distinguish waves of different energy.

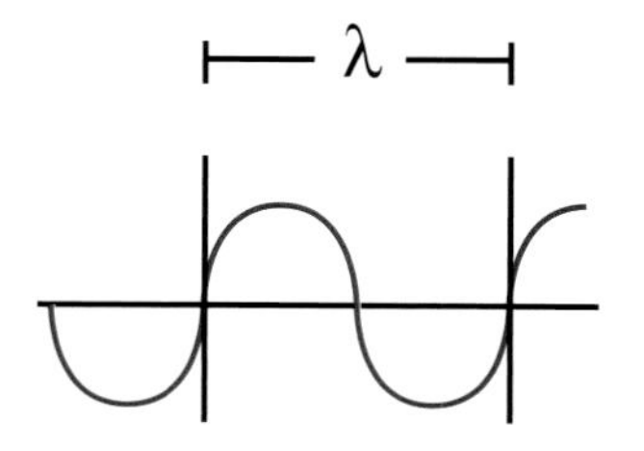

Wavelength	λ (m)	Distance between peaks of the wave
Period	T (sec)	Time to travel one λ or cycle of the wave
Frequency	f (Hz)	The inverse of the period. Cycles passing a point per second. 1 cycle = 2π radians $f = 1/T$
Angular frequency	ω (rad/s)	$\omega = 2\pi/T = 2\pi f$
Wave amplitude	A	Height of wave above the average position
Speed	v (m/s)	Linear velocity of the wave $v = \lambda \times f$
Wave number	k (m^{-1})	$k = 2\pi/\lambda$

Sound Waves

- **Wave Nature of Sound:** Sound is a compression wave that displaces the medium carrying the wave. Therefore, sound cannot travel through a vacuum.

- **General Speed of Sound:**

$$v=\sqrt{\frac{B}{\rho}}$$

B is the **bulk modulus,** the volume compressibility of the solid, liquid or gas.

ρ is the density of the material.

- Sound travels more quickly in denser, rigid materials than it does in less-dense compressible materials. (Consider rubber and copper.)

- **For a gas:**

$$v=\sqrt{\frac{\gamma RT}{M}}$$

NOTES:

$\gamma = C_p/C_v$ (the ratio of gas heat capacities).

- ◆ This quantifies how sound travels in a gas composed of particles moving at an average speed of $\sqrt{\frac{3RT}{M}}$

Table of Sample Values (in m/s)	
Gases	
Helium (0°C)	965
Hydrogen (27°C)	1310
Nitrogen (27°C)	353
Oxygen (27°C)	330
Air (25°C) (80/20 N_2/O_2)	346
Liquids	
Liquid water	1493
Methanol	114
Solids	
Aluminum	5100
Copper	3560
Lead	1322
Rubber	54

- **Loudness – Intensity & Relative Intensity:** Loudness or sound intensity is the power carried by a sound wave.

  Absolute Intensity (I = Power/Area) is an inconvenient measure of loudness.

  Relative Loudness: The Decibel scale **(dB)**

  $$\beta(\mathbf{dB}) = \mathbf{10\ log}\left(\frac{\mathbf{I}}{\mathbf{I_0}}\right)$$

  The Decibel scale is defined relative to the threshold of hearing, I_0. (The minimum on this scale "0"; $\beta(I_0) = 0$ dB.)

Intensity of Various Sounds (in dB)	
Rocket engine	200
Jet plane	150
Tractor	120
Pneumatic drill	100
Vacuum cleaner	85
Heavy traffic	75
Conversation	65
Whisper	40
Threshold of hearing	0

A change in 10 dB represents a 10 fold increase in sound intensity, I.

■ **Doppler Effect:** The sound frequency shifts **(f '/ f)** due to relative motion of the source of the sound and the observer or listener.

$\mathbf{v_0}$ – speed of the listener or observer
$\mathbf{v_s}$ – speed of the sound source
$\mathbf{v}$ – speed of sound

Case #1: If the source of sound is approaching the observer, the frequency increases.

$$\frac{f'}{f} = \left(\frac{v + v_0}{v - v_s}\right)$$

O => <= S

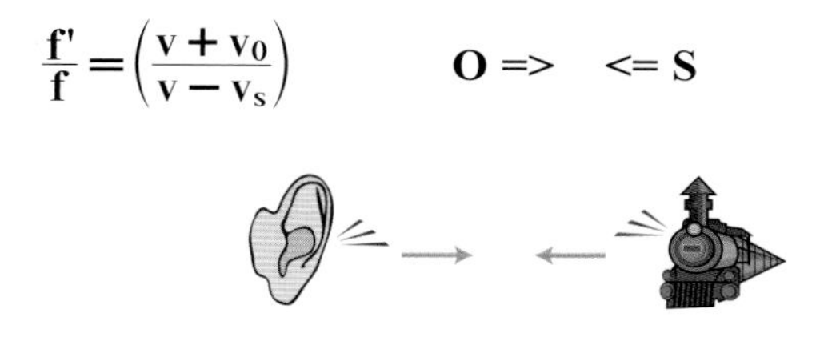

Case #2: If the source of sound is moving away from the observer, the frequency decreases.

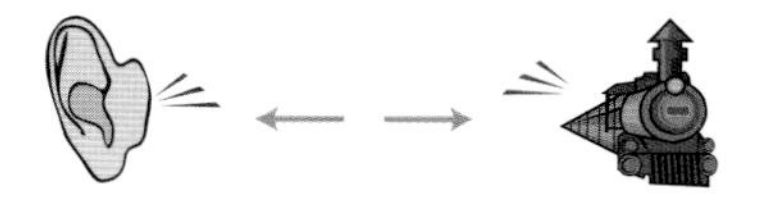

$$\frac{f'}{f} = \left(\frac{v - v_0}{v + v_s}\right)$$

<= O S =>

The key is to identify the relative speed of the source and listener

For $v_0 = 0$ and $v_s = 0$, f'/f = 1, no Doppler effect.

Thermodynamics

Goal: Study of the work, heat and energy associated with a physical or chemical process.

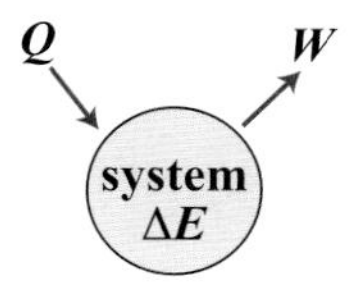

KEY Variables	
Heat: Q	+Q added to the system
Work: W	+W done by the system
Energy: E	System internal Energy
Enthalpy: H	$H = E + PV$
Entropy: S	Thermal disorder
Temperature: T	Measure of thermal Energy
Pressure: P	Force exerted by a gas
Volume: V	Space occupied

Thermodynamic variables are of two types. A **variable of state** is independent of the path or steps in a process.

Examples include **temperature, pressure, volume, energy, enthalpy** and **entropy.** The change in energy for a process does not depend on how the work is performed. In addition, measuring the temperature of a sample tells you nothing about the history of the sample.

The other group includes experimental variables that are state or **path-dependent,** such as **work** and **heat.** For example, the work generated by a motor depends on how the motor is cooled.

Types of Processes

Experimental conditions can be controlled to allow for different types of processes.

These choices constrain the behavior of thermodynamic variables, ΔE, Q, W, T, V and P.

Condition	Constraints	Thermodynamic Result
Isothermal	$\Delta T = 0$ Fixed Temperature	$\Delta E = 0$, $Q = W$ $PV = \text{constant}$ Heat flow matches the work
Adiabatic	$Q = 0$ No heat flow	$\Delta E = -W$ $PV^{\gamma} = \text{constant}$ Internal energy is expended to perform work

Condition	Constraints	Thermodynamic Result *(continued)*
Isobaric	$\Delta P = 0$ Fixed pressure	$W = P \Delta V$ $\Delta H = Q$
Isochoric	$\Delta V = 0$ Fixed volume	$\Delta E = Q$ $W = 0$ Work is typically tied to $P \Delta V$

Temperature & Thermal Energy

Temperature is a measure of thermal energy of a body. A warmer object possesses more thermal energy than a cooler object. Heat naturally flows from a hotter to a cooler object.

- Temperature is measured **in Kelvin, *absolute temperature:***

$$T(K) = T(°C) + 273.15$$

NOTES:

- **T(K) is always *positive.*** Lab temperature is often measured in °C or °F, and must be converted to Kelvin for any calculations.
- **T always refers to Kelvin,** unless specifically noted in the equation.

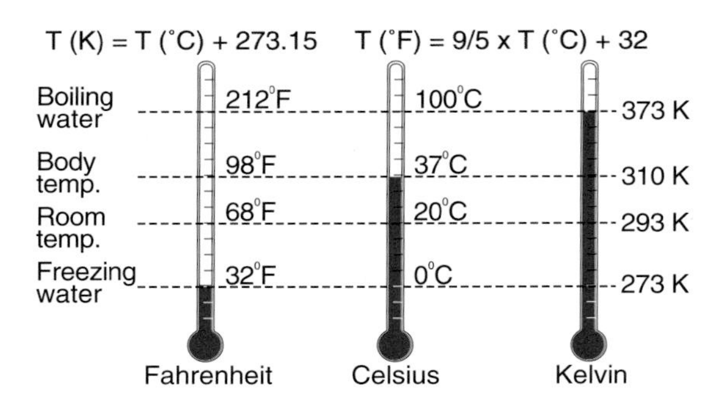

- **Zeroth Law of Thermodynamics:** By definition, if two bodies are in thermal equilibrium, they are at the same temperature. If two bodies, #1 and #2, are separately in thermal equilibrium with a third body, #3, then bodies #1 and #2 are also in thermal equilibrium. This is similar to the commutative relationships in algebra.

Thermal Expansion

Goal: Determine the change in the length (L) or volume (V) as a function of temperature. Applies to a solid, liquid or gas.

- **Solid:** $\frac{\Delta L}{L} = \alpha \Delta T$

  The relative change in length is proportional to a change in temperature. For metals, a large change in temperature is required to give a noticeable change in length.

For example, a ΔT of 1,000°C changes the length by 1%.

Liquid: $\frac{\Delta V}{V} = \beta \Delta T$

The relative change in volume is proportional to a change in temperature.

β has the units of inverse temperature, usually $°C^{-1}$.

For the same material, the volume thermal expansion factor, β, is three times the linear thermal expansion factor, α:

$\beta = 3\alpha$

Gas: $\Delta V = \frac{(T_2 - T_1)\, nR}{P}$

This is Charles' Law; volume is proportional to absolute temperature:

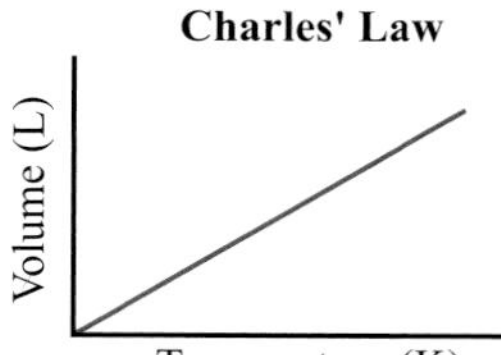

Examples of Thermal Expansion α & β	
β for several liquids	
Water	$0.21 \times 10^{-3}\ °C^{-1}$
Methanol	$1.49 \times 10^{-3}\ °C^{-1}$
Ethanol	$1.4 \times 10^{-3}\ °C^{-1}$
Mercury	$1.81 \times 10^{-3}\ °C^{-1}$

Examples of Thermal Expansion α & β *(continued)*	
α for several metals	
Brass	20.3 x 10^{-6} °C^{-1}
Aluminum	23.0 x 10^{-6} °C^{-1}
Nickel	13.3 x 10^{-6} °C^{-1}
Zinc	32.5 x 10^{-6} °C^{-1}

Heat Capacity

Goal: Determine how the heat content and temperature are related.

If you add heat to a body, the temperature will increase. Conversely, a change in temperature indicates a change in heat content for a body.

- The heat capacity, C, quantifies the relationship between a temperature change, del T, and the heat lost or gained, Q.

  $$C = \frac{Q}{\Delta T}$$

  or

  $$Q = C\Delta T$$

  It is useful to standardize this as the **specific heat capacity,** defined as the heat capacity per mass.

- C may also be defined as a **molar heat capacity.**

  For the molar heat capacity, the unit of C is:
  J/mol K

- Two special experimental cases: **Fixed Pressure or Fixed Volume.**

  Heat capacity for constant pressure: C_p

  Typical lab conditions – constant atmospheric pressure

  Enthalpy is the key variable.

  Heat capacity for constant volume: C_v

  Energy is the key variable.

Sample Molar C_p Values (J/mol K)	
Aluminum	24.25
Boron	11.4
Chromium	23.47
Carbon (graphite)	8.58
Water	75.3

- **For an Ideal Gas:**

  $C_p = \frac{5}{2} R$

  and

  $C_v = \frac{3}{2} R$

The **ratio** of these two heat capacities is called γ:

$$\gamma = \frac{C_p}{C_v} = \frac{5}{3} = 1.667$$

- **Carnot's Law:** For ideal gas: $C_p - C_v = R$
For experiments with fixed **volume,** use ΔE to describe the thermal properties of the system:

$$\Delta E = C_v \Delta T$$

This equation applies for a constant **volume** process.

For experiments with fixed **pressure,** use ΔH to describe the thermal properties of the system:

$$\Delta H = C_p \Delta T$$

This equation applies for a constant **pressure** process.

NOTES:
Carnot's Law is exact for monatomic gases; it must be modified for molecular gases. A molecule, such as water, will have vibrational and rotational motion and energy terms associated with this motion. These terms will also contribute to the heat capacity.

Ideal Gas Law

Goal: Derive and use a simple equation of state for a gas

PV = n RT

The **Ideal Gas Law** is an **idealized equation** of state for a gas. It is based on **massless, non-interacting** gas particles.

Key Variables

- **Pressure, P:** The standard unit is the Pascal, (Pa), but more commonly, the bar is used as the unit of pressure.

  1 bar = 10^5 Pa

  Other work uses the standard atmosphere, atm:

  1 atm = 9.8 x 10^5 Pa = 0.98 bar

- **Volume, V:** The standard unit is the m^3, but in practice, the liter, L, is more common.

  1 L = 1 dm^3; 1 m^3 = 1,000 L

- **Temperature, T:** The standard temperature unit is absolute temperature based on the Kelvin scale: **T(K)**

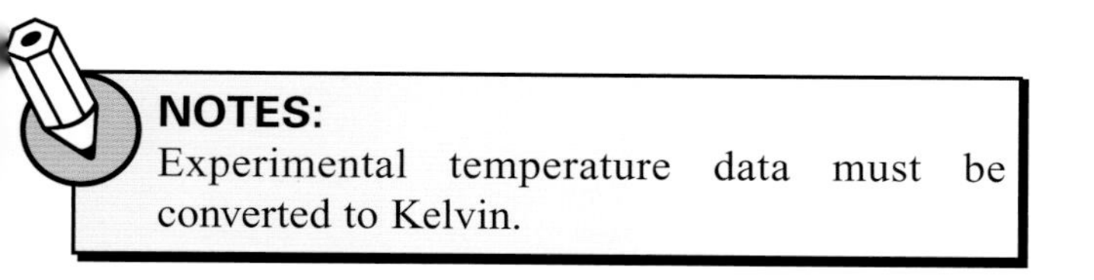

- **Amount of gas, n:** Given by the number of moles of gas (mol), symbol n.
- **R is a proportionality constant,** called the gas constant, given the symbol R:

$R = 8.314\ J\ mol^{-1}\ K^{-1}$

For gas calculations, R is modified to fit the units of the Gas Law:

$R = 0.083\ L\ bar\ mol^{-1}\ K^{-1}$
For pressure in units of "atm":

$R = 0.082\ L\ atm\ mol^{-1}\ K^{-1}$

⚠**PITFALL:** It is easy to make errors in the units of R, T, P or V.

Key Applications

NOTES:
The Gas Laws can be used in various experimental conditions.

- **Boyle's Law** (Constant Temperature)

Pressure is proportional to 1/Volume, with Temperature fixed.

$$P \propto \frac{1}{V}$$

or

$P \times V$ = constant

- Pressure is proportional to temperature, with volume fixed.

  $P \propto T$

- **Charles' Law** (Constant Pressure)

Volume is proportional to temperature, with Pressure fixed.

  $V \propto T$

- **Avogadro's Law**

  Volume is proportional to the number of moles, n.

  $V \propto n$

- **General Ideal Gas Law Application:**

Use PV = nRT to examine a gas sample under specific conditions of Pressure, Volume, number of moles and Temperature.

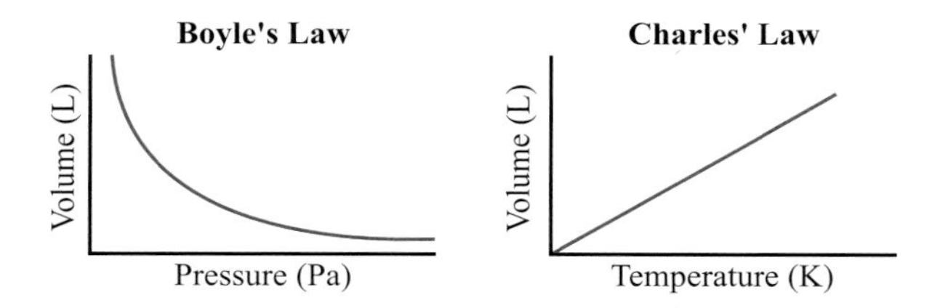

Enthalpy & the First Law of Thermodynamics

Goal: Determine Q, ΔE and W for a process.

For any process, the work, W, and heat, Q, depend on path of the process. However, in all cases, the change in energy, ΔE, a state variable, is independent of path or process.

Guiding Principle:

- First Law of Thermodynamics:

  $$\Delta E = Q - W$$

Key Concept: Conservation of the system's energy in the process.

- The change in **energy** of the system is determined by the difference between the **heat** *gained* by the system and the **work** *performed* by the system on the mechanical surrounding.

Applications of the First Law involve an examination of changes in the temperature, pressure, work and heat of the system or process.

- **Enthalpy:** A new state variable obtained from applying the First Law of Thermodynamics at constant pressure.

  $$H = E + PV$$
  $$\Delta H = \Delta E + P\,\Delta V$$

- **$\Delta H = Q$** for a process at constant pressure (i.e., $\Delta P = 0$).

- Constant pressure is the typical condition for most experiments in the laboratory.

A process with a **positive ΔH** is called an **endothermic** process. ***The system absorbs heat from the surroundings.***

NOTES:
Evaporation of liquid to gas and **melting** of a solid both are ***endothermic*** processes.

A process with a **negative ΔH** is called an **exothermic** process. ***The system releases heat to the surroundings.***

NOTES:
Combustion of fuel and **condensation** of vapor to liquid both are ***exothermic*** processes.

Phase Transitions

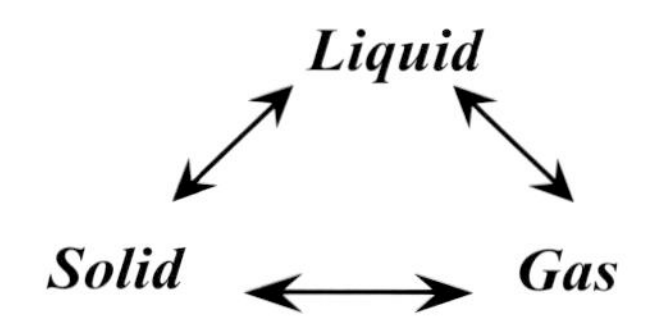

- A change of phase corresponds to a change in **enthalpy.**

- Vaporization of a liquid to a gas requires the addition of heat, called the **ΔH of vaporization:**

  ΔH_{vap}

- The melting of a solid, forming a liquid, has a **ΔH of fusion:**

  ΔH_{fus}

- In order to melt a solid, ΔH_{fus} must be added to the solid.

Variable Temperature

$$\Delta H = \int C_p \, dT$$

For constant C_p:

$$\Delta H = C_p \, \Delta T$$

The difference between E and H is the **work** performed by the process.

Examples of work include:

$$W = \int P dV$$

P is the pressure that opposes the volume change for an expansion, or the pressure that causes the volume change for gas compression.

- W is not a state variable; it depends on the path of the process.
- The maximum work is performed for an idealized, reversible process.

◆ This process is called an **isobaric expansion.**

◆ **Reversible,** isothermal expansion of an Ideal Gas against P_{ext}.

◆ Gas expands from V_1 to V_2 using an infinite number of steps; the system is in equilibrium at all times:

$$W = n\,RT \ln\left(\frac{V_2}{V_1}\right)$$

◆ The same expression works for compression, with V_1 and V_2 interchanged.

◆ **Single step** expansion from V_1 to V_2 against an opposing pressure, P_{ext}. At the end, the pressure of the gas is P_{ext}:

$$W = (V_2 - V_1)\,P_{ext} = \Delta V\,P_{ext}$$

Work Performed by a Single Step Gas Expansion

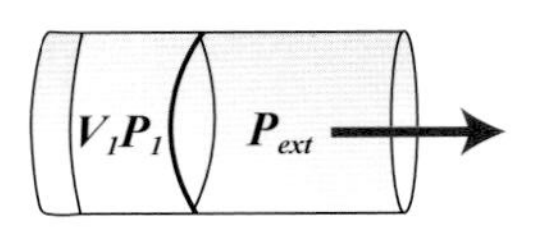

Before Expansion

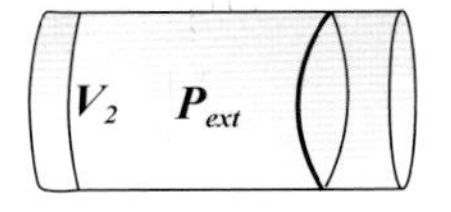

After Expansion

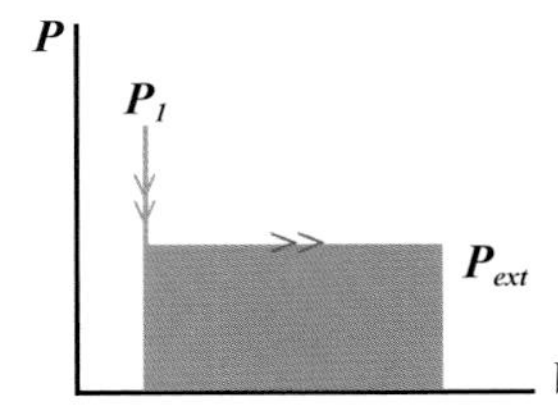

Work Performed

The Kinetic Theory of Gases

- Gas molecules are in constant motion, exerting pressure and having kinetic energy.
- The kinetic energy can be examined in terms of:

  - particles in motion, $\mathbf{E_k = {}^1/_2\, mv^2}$

  - properties of an Ideal Gas that describes the energy of a mole of gas particles as $\mathbf{\frac{3}{2}RT}$

Key Equations for Energy of an Ideal Gas

$$E = {}^1/_2\, mv^2$$

and

$$E = \frac{3}{2}RT$$

- **Average Speed of a Gas Molecule:** Equating these expressions gives an estimate of the "average" speed, called the **root-mean-square speed, v_{rms}:**

$$v_{rms} = \sqrt{\frac{3RT}{M}}$$

Examples of v_{rms} for Various Gases at 20°C		
Gas	v_{rms} (m/s)	M (g/mol)
H_2	1900	2.0
He	1350	4.00
Water	640	18.0
CO_2	410	44.0

- **Gas Speed & Temperature:** Gas speed, v_{rms}, is proportional to $\sqrt{T}$; therefore, changing the temperature from T_1 to T_2 changes the speed by a factor of $\sqrt{\frac{T_2}{T_1}}$.

NOTES:
T is in Kelvin, not Centigrade or Fahrenheit.

- **Gas Speed & Mass:** $\mathbf{v}_{\mathbf{rms}}$ is proportional to $\sqrt{\frac{1}{M}}$.

  Larger molecules move more slowly than smaller molecules.

- **Kinetic Energy for 1.00 Mole of an Ideal Gas**

  $\frac{3}{2}RT$

NOTES:
T, an Ideal Gas, is composed of point particles.

- **For a Real Gas:** Heat capacity and energy terms must be added for the vibrational and rotational degrees of freedom of the molecule.

Entropy & the Second Law of Thermodynamics

- The Second Law of Thermodynamics is concerned with the **driving** forces behind physical and chemical processes.
- Every system naturally proceeds to equilibrium state.
  - For example, a hot object naturally cools down to the surrounding temperature.

Goal: Examine the driving force for processes.

Key Variables

- **Entropy: S**

  Entropy measures the thermal disorder of a system.

$$dS = \frac{dQ}{T}$$

Entropy is a state variable, like energy and enthalpy.

- **S(univ) = S(system) + S(thermal reservoir)**

ΔS(univ) = ΔS(system) + ΔS(thermal reservoir)

Guiding Principle: Second Law of Thermodynamics

- For any process, $\mathbf{\Delta S_{univ} > 0}$.
- One exception: $\mathbf{\Delta S_{univ} = 0}$ for a reversible process.
- This Law defines a system at equilibrium and describes the natural driving forces for spontaneous processes.

Examples of Entropy Changes

- **Natural Heat Flow:** Heat flows from T_{hot} to T_{cold}

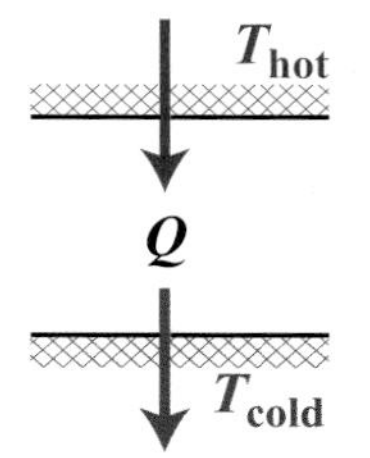

- **Entropy & Phase Changes**

$$\Delta S\,(\text{phase change}) = \frac{\Delta H\,(\text{phase change})}{T\,(\text{phase change})}$$

solid → liquid	**positive ΔS**	**ΔS_{fusion}**
liquid → gas	**positive ΔS**	**$\Delta S_{vaporization}$**

- A gas has more entropy than a liquid, which has more entropy than a solid.
- **Entropy** reflects randomness and disorder in the system.
- **Entropy & Temperature for an Ideal Gas**

$$S(T): \Delta S = n\, Cp \ln\left(\frac{T_2}{T_1}\right)$$

- Increasing the temperature also increases the disorder of a material.
- This principle also applies to liquids and solids, though the mathematical form may differ.

- **Entropy & Volume for an Ideal Gas**

- As gas expands from V_1 to V_2, the entropy change is given by:

$$S(V): \Delta S = n\, R \ln\left(\frac{V_2}{V_1}\right)$$

- The disorder of a gas increases if it fills a larger volume.
- This property explains why gases freely mix: Mixing is spontaneous because gases freely expand to fill their container(s).

Heat Engines

Goal: Examine Q and W of a heat engine.

Thermal Engine: The heat engine transfers heat, Q, from a hot to a cold reservoir, and produces work in the process, denoted by W.

Thermal engine

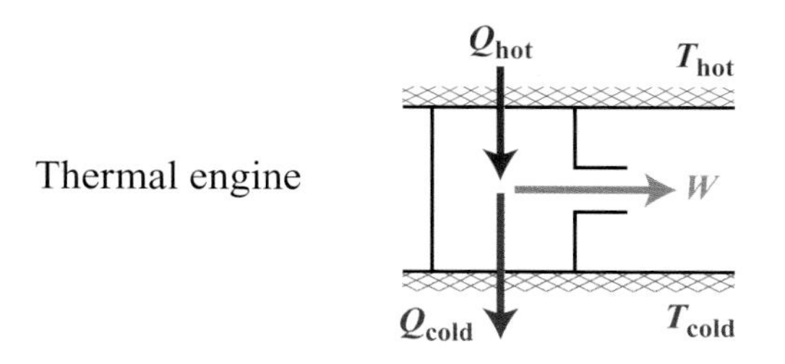

- The First Law of Thermodynamics states that the work, W, must equal the difference between the heat terms:

$$\mathbf{W = Q_{hot} - Q_{cold}}$$

- The efficiency of an engine is defined as the ratio of work done by the engine, W, divided by the heat extracted from the hot reservoir, Q_{hot}.
- The symbol η is used for engine efficiency:

$$\eta = \frac{W}{Q_{hot}}$$

or

$$\eta = 1 - \frac{Q_{cold}}{Q_{hot}}$$

Idealized Heat Engine: The Carnot Cycle

- The **Carnot Cycle** consists of four reversible steps that form a cycle:

  - two **isothermal** steps
  - two **adiabatic** steps

Refer to figure below:

Step 1: isothermal expansion at T_{hot}	A to B
Step 2: adiabatic expansion T_{hot} to T_{cold}	B to C
Step 3: isothermal compression at T_{cold}	C to D
Step 4: adiabatic compression T_{cold} to T_{hot}	D to A

For overall cycle: **$\Delta T = 0$, $\Delta H = 0$ and $\Delta S = 0$**

- **Thermal Efficiency of the Carnot Engine:**

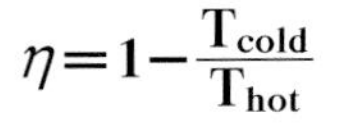

$$\eta = 1 - \frac{T_{cold}}{T_{hot}}$$

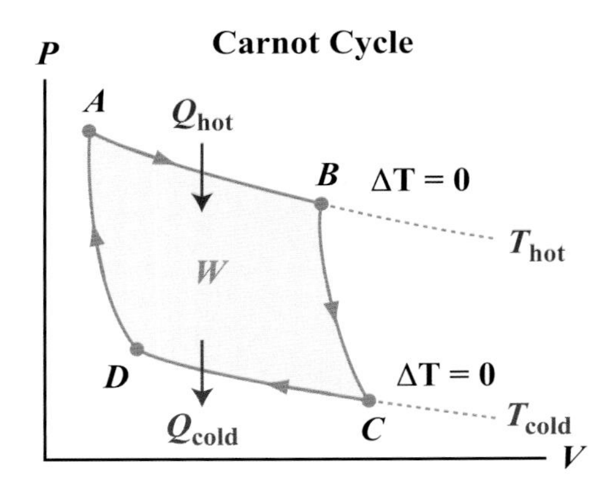

Electricity & Magnetism

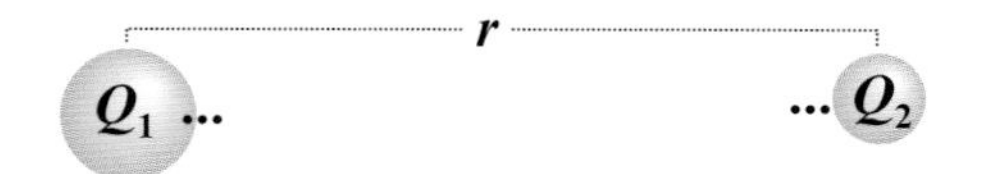

Electric Fields & Electric Charge

Goal: Examine the nature of the field generated by an electric charge and the forces between charges.

Key Variables & Equations

- **Coulomb,** given the symbol C, is a measure of the amount of charge.
- Coulomb is defined as "1 ampere sec" of charge.

1 amp = 1 Coulomb/sec

- **e** (small e) is the charge of a single electron.

$$e = 1.6022 \times 10^{-19}\ C$$

- **Coulomb's Law** describes the electrostatic force between charge q_1 and q_2 separated by a distance r:

$$F_{coul} = \frac{1}{4\pi\varepsilon_0}\frac{q_1 q_2}{r_2}\hat{r}$$

- The unit vector **r** is defined by charge orientations; it is along the line connecting the two charges, q_1 and q_2.
- **Units for Force Calculations**

$$F = 9 \times 10^9 \, N \frac{q_1(C)\, q_2(C)}{r\,(m)^2}$$

NOTES:
q must be in Coulombs and must be in meters.

- **Electric Field, E,** is the potential field generated by a charged body.
- It is defined by the force, F, produced by the field, E, interacting with a test charge q_0.

$$E = \frac{F}{q_0}$$

- **Superposition Principle:** Forces and fields are composites of contributions from each charge in the system.

$$F = \sum F_i$$

$$E = \sum E_i$$

Hint: Forces and Electric Fields are vectors. The sum must account for the *x, y* and *z*-components for the force and electric field.

Sources of Electric Fields: Gauss's Law

Goal: Define electric flux, φ_e

Key Variables & Equations

■ **Gauss's Law**

$$\varphi_e = \int E \times dA = \frac{Q}{\varepsilon_0}$$

■ The electric flux, φ_e, depends on the **total charge** in the closed region of interest.

Electric Potential & Coulombic Energy

Goal: Determine **Coulombic** potential energy for charged bodies.

Key Variables & Equations

■ Coulombic potential energy is derived from Coulombic force using the following equation:

$$U_{coul} = \int F_{coul}\, dr$$

■ **Coulombic Potential Energy:**

$$U_{coul} = \frac{1}{4\pi\varepsilon_0} \frac{qq'}{r}$$

■ **Coulombic Potential:** The Coulomb potential, V(q), generated by charge q is obtained by dividing the Ucoul by the test charge, q'.

$$V(q)=\frac{U}{q'}=\frac{1}{4\pi\varepsilon_0}\frac{q}{r}$$

$$U=V(q)q'$$

NOTES:
The potential is scalar, depending on |r|.

■ For an array of charges, q_i:

$$V_{total}=\sum V_i$$

■ **Units for U(r):**

$$U=9.0\times10^9\,J\,\frac{q_1(C)q_2(C)}{r(m)}$$

NOTES:
q must be in Coulombs and r must be in meters.

■ **Potential for a Continuous Charge Distribution:**

$$V=\frac{1}{4\pi\varepsilon_0}\int\frac{dq}{r}$$

Dielectric Properties

- **The Dielectric Effect:** Electrostatic forces and energies are diminished or screened by the presence of matter with **dielectric constant, κ**, between the charges.
- Voltage and electrostatic force (V & F) depend on the dielectric constant, κ.
- Equations for interactions in a vacuum can be modified to treat interactions in other media by replacing ε_0 with $\kappa\,\varepsilon_0$; κ is the dielectric constant for the material.
- For a material with dielectric constant, κ:

$$V(\kappa)=\frac{1}{\kappa}V\,(\textbf{vacuum})$$

$$F(\kappa)=\frac{1}{\kappa}F\,(\textbf{vacuum})$$

Sample Values of Dielectric Constants		
Material	Dielectric Constant κ	Dielectric Strength (kV/mm)
Vacuum	1	Infinity
Air	1.00059	3.0
Polypropylene	2.3	23.6
Polyethylene	2.3	18.9
Water	80	65–70
Natural rubber	2.6	100–200
Butyl rubber	2.5	23.6

- **Dielectric Strength:** The maximum voltage that can be applied to a capacitor filled with this material.
- A vacuum has a dielectric strength of infinity. This parameter determines the maximum operating voltage for a capacitor.
- Beyond this voltage, the material begins to conduct electricity, and the capacitor breaks down.
- The unit is V/m or more practically, kV/mm.

Capacitance & Dielectrics

- A **capacitor** consists of two electrical conducting plates carrying equal and opposite charges separated by some distance.
- A capacitor is a device used to store charge or electrical potential energy.
- The properties depend on the charge, Q, the distance between the plates, the area of the plates, the voltage placed on the device, and the material between the plates.

Key Equations

- **Capacitance,** C, is defined as the ratio of charge, Q, divided by the voltage, V, for a capacitor.

$$C=\frac{Q}{V}$$

- V is the measured voltage.
- Q is the charge.
- **Energy stored** in a charged capacitor:

$$U=\frac{\frac{1}{2}Q^2}{C}=\frac{1}{2}QV=\frac{1}{2}CV^2$$

■ **Parallel plate** capacitor, with a **vacuum,** with area A, and spacing d:

Capacitance $C = \varepsilon_0 \frac{A}{d}$

Energy Stored $U = \frac{1}{2} \varepsilon_0 A d E^2$

Electric Field $E = \frac{V}{d} = \frac{Q}{\varepsilon_0 A}$

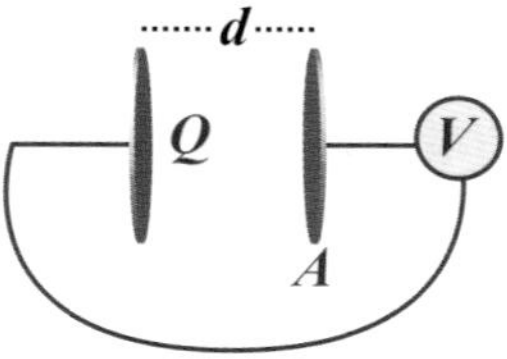

■ Parallel plate capacitor, dielectric material with dielectric constant **κ,** with area A, spacing d:

$$C = \frac{\kappa \varepsilon_0 A}{d} = \kappa C_0$$

C_0: vacuum capacitor

■ **Capacitors in Circuits:** Groups of capacitors in circuits can be shown to behave equivalently to a single capacitor with properties calculated from the separate capacitors.

■ **Series Combinations:** For capacitors in series, the voltage drop across each capacitor must add up to the voltage drop across a single capacitor:

$$V_{tot} = V_1 + V_2$$

- Therefore, the inverse of the capacitance of the total is found to equal the sum of the inverse of each capacitance:

$$\frac{1}{C_{tot}} = \frac{1}{C^1} + \frac{1}{C^2} + \frac{1}{C^3} \cdots$$

- **This applies to any number of capacitors in series.**
- **Capacitors in Series:**

$$\frac{1}{C_{tot}} = \Sigma \frac{1}{C_i}$$

- **Parallel Combinations:** The total charge held by parallel capacitors is simply the sum of the charge on each capacitor. Therefore, the total capacitance, C_{tot}, of two capacitors in parallel, with capacitance, C_1 and C_2, is $C_1 + C_2$.
- **This applies to any number of capacitors in parallel.**
- **Capacitors in Parallel:**

$$C_{tot} = \Sigma C_i$$

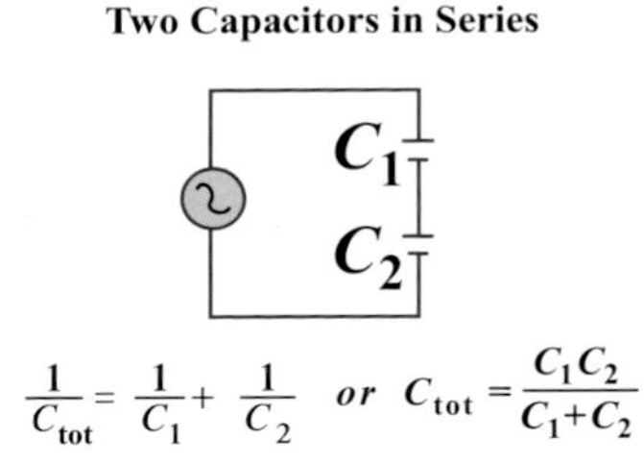

Two Capacitors in Parallel

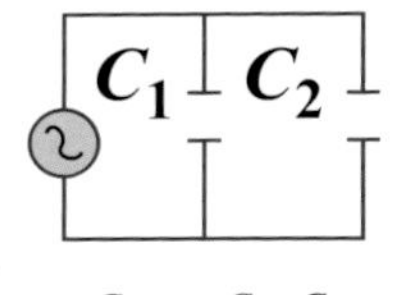

$$C_{tot} = C_1 + C_2$$

Current & Resistance of a Conductor

- **Current & Charge:** The current, I, measures the amount of charge passing through a conductor over a time interval, t.
- The quantity of charge, Q, is given by the product of current × time.

Total charge, Q: $\mathbf{Q = I \times t}$

- **Ohm's Law:** Current flowing through a conductor produces an electric field, E, that is proportional to the current density, J.
- The proportionality constant, denoted **σ,** is called the **conductivity.**

$$\mathbf{J = \sigma E}$$

- **Resistance:** The resistance, R, measures the fact that conduction is not ideal; some energy is lost in the process.
- Resistance is defined as the voltage divided by the current:

$$R = \frac{V}{I}$$

- The SI unit for resistance is the Ohm, given the symbol Ω.

$$\mathbf{1\ ohm\ (\Omega) = \frac{1\ volt\ (V)}{1\ amp\ (A)}}$$

■ **Resistivity:** The inverse of the conductivity of a material is called the resistivity, given the symbol **ρ:**

$$\rho=\frac{1}{\sigma}$$

■ **Voltage** for current flowing in a conductor.
■ The voltage is given by the product of the current × resistance.
■ Voltage is symbolized V:

$$V = IR$$

■ **Power Dissipation:** Since the conduction is not ideal, power is dissipated as the current passes through the resistive conductor.
■ Power dissipated is given by the product of voltage and resistance, or the resistance times the square of the current:

Power Dissipation, P:

$$P = VR = I^2 R$$

■ **Resistors in Circuits:** Certain groups of resistors in a circuit are found to behave as a single resistor, noted as R_{tot}.
■ For **Resistors in Series:** Combining resistors in series simply increases the resistance of the circuit.
■ So the total resistance is simply the sum of the individual resistors:

$$R_{tot}=\sum R_i$$

- For **Resistors in Parallel:** Combining resistors in parallel does not give a simple additive effect.
- In this case the current, I, must be additive, and the voltage is the same for each of the resistors in the circuit.
- This results in the following equivalence:

$$I_{tot} = \frac{V}{R_{tot}} = I_1 + I_2 = \frac{V}{R_1} + \frac{V}{R_2}$$

- Or, for resistance:

$$\frac{1}{R_{tot}} = \sum \frac{1}{R_i}$$

Two Resistors in Series

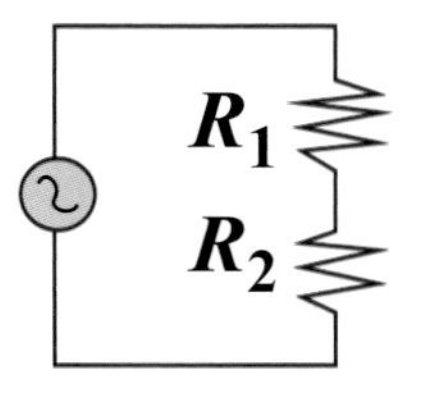

$$R_{tot} = R_1 + R_2$$

Two Resistors in Parallel

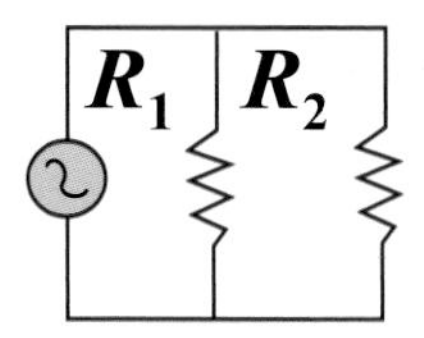

$$\frac{1}{R_{tot}} = \frac{1}{R_1} + \frac{1}{R_2}$$

$$or \quad R_{tot} = \frac{R_1 R_2}{R_1 + R_2}$$

Sample Resistivities (Ω m)		
Silver	1.6×10^{-8}	**best conductor**
Copper	1.7×10^{-8}	
Gold	2.5×10^{-8}	
Aluminum	2.8×10^{-8}	
Carbon	3.5×10^{-5}	
Glass	1×10^{12} (varies)	
Quartz	8×10^{17}	**worst conductor**

Direct Current (DC) Circuits

Goal: Examine a circuit containing a battery, resistors and capacitors; determine voltage and current properties.

Key Equations & Concepts

- **emf:** The voltage of a circuit is called the **electromotive force,** denoted **emf.**
- This voltage accounts for the battery, V_b, and the circuit voltage, denoted by IR.

$$\mathbf{emf = V_b + IR}$$

- The battery also has an internal resistance, denoted r:

$$r = \frac{V_b}{I}$$

Circuit Terminology

- **Junction:** a connection of three or more conductors
- **Loop:** a closed conductor path
- Replace resistors in series or parallel with R_{tot}
- Replace capacitors in series or parallel with C_{tot}

Kirchoff's Circuit Rules

- **Constraints on the Voltage:** For any loop in the circuit the voltage must be the same:

$$\sum V = \sum IR$$

- The energy must be conserved in the circuit loop.

Constraints on the Current

- The current must balance at every node or junction.
- For any junction:

$$\sum I = 0$$

- The total charge must be conserved in the circuit; the amount of charge entering and leaving any point in the circuit must be equal.

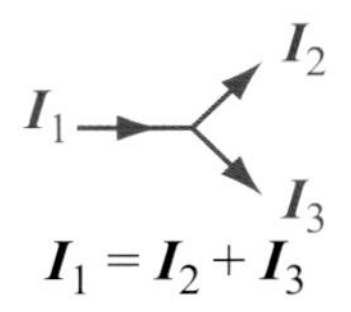

Magnetic Fields

Key Concepts

- **Magnetic Field:** A moving electric charge or current generates a **magnetic field,** denoted by the symbol **B.**
- The vector B is also called the **magnetic induction** or the **magnetic flux density.**
- The **SI unit** for a ***magnetic field*** is the **Tesla, T.**
- The **SI unit** for ***magnetic flux*** is the **Weber, Wb.**
- **Tesla** is defined as 1 Weber/m^2; the magnetic field strength that generates a force of 1 N on 1 coulomb of charge moving at 1 m/s:

$$1\,\mathrm{T} = \frac{\mathrm{Wb}}{\mathrm{m}^2} = \frac{\mathrm{N}}{\mathrm{C}} \times \frac{\mathrm{m}}{\mathrm{s}} = \frac{\mathrm{N}}{\mathrm{A}} \times \mathrm{m}$$

- The CGS unit is the **Gauss, G:**

$$1\ \mathrm{T} = 10^4\ \mathrm{G}$$

- For a **bar magnet,** the field is generated from the ferromagnetic properties of the metal forming the magnet.
- The poles of the magnet are denoted North/South.

- The field lines are shown in figure below.

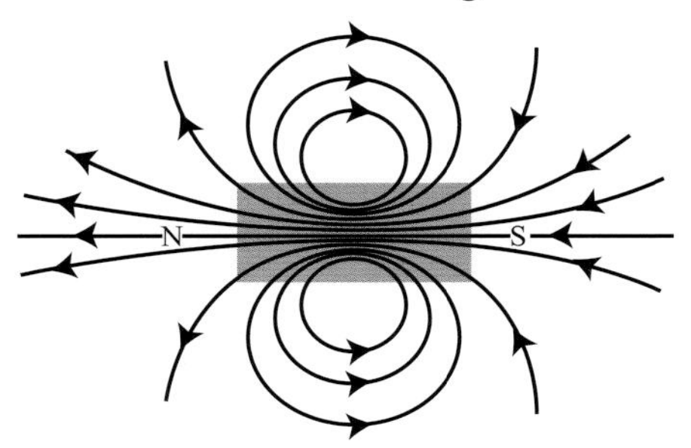

- For a **current loop**, the field is generated by the motion of the charged particles in the current.
- The field lines for such a current loop are shown in figure below.

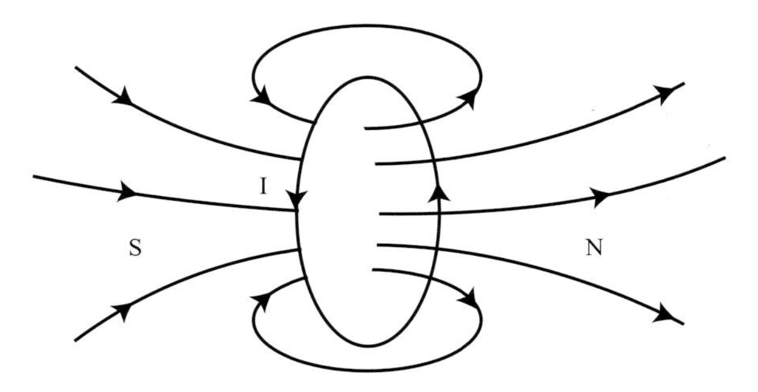

- **Magnetic Force:** The force on charge **q** moving at velocity **v** in magnetic field **B** is given by the following expression:

$$\mathbf{F} = \mathbf{q}\ \mathbf{v} \times \mathbf{B} = \mathbf{qvB}\ \sin\boldsymbol{\theta}$$

- **θ** is the angle between v and B (both are vectors).
- This expression involves a vector cross product.

- **For v parallel to B:**

  $F = 0$ ($\theta = 0$, minimum force)

- **For v perpendicular to B:**

  $F = q\,v\,B$ ($\theta = \pi/2$, maximum force)

- **The "right hand rule" defines the direction of the force.**

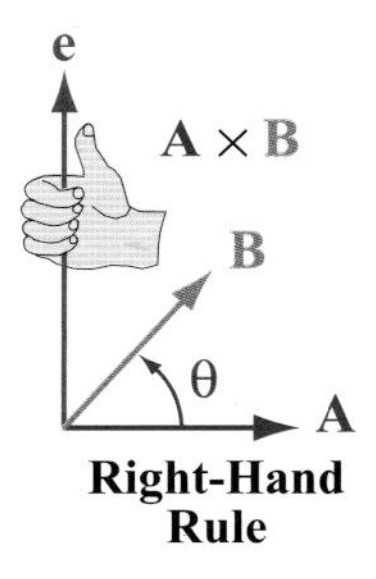

Right-Hand Rule

- **Force on a conducting segment:** For a current I passing through a conductor of length l in a magnetic field B, the force is given by:

  $$\mathbf{F} = \mathbf{I\,l \times B}$$

- For a general current path s:

  $$F = I\int ds \times B$$

- For a **closed** current loop:

  $$\mathbf{F} = \mathbf{0}$$

- **Magnetic Moment:** A **magnetic moment,** denoted M, is produced by a current loop. Some materials generate a moment due to electronic properties, such as ferromagnetism.
- A current loop, with current I and area A, generates a magnetic moment of strength M:

$$\mathbf{M = I\,A}$$

- **Torque on a Loop:** A loop placed in a magnetic field will experience a torque, rotating the loop.
- The torque on a loop, denoted $\boldsymbol{\tau}$**,** depends on the cross product of M and B:

$$\boldsymbol{\tau} = \mathbf{M \times B}$$

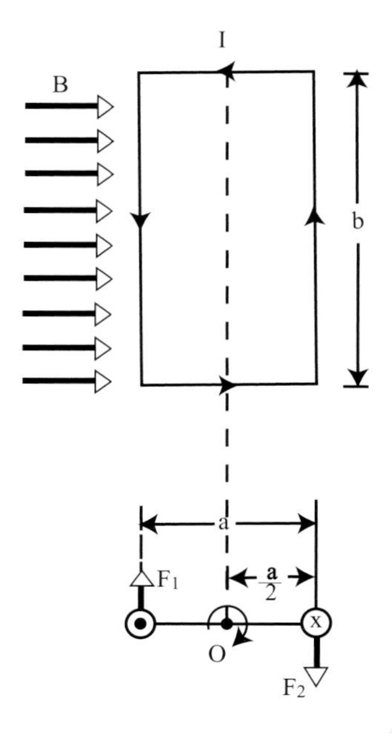

- **Magnetic Potential Energy:** The energy of interaction between a magnetic field and magnetic moment is given from the dot product of these two vectoral properties:
  U(magnetic) = - M • B

- **Lorentz Force:** A charge interacts with both E and B; the force is given by the following expression:
  F = q E + q v x B

> **NOTES:**
> Both **B** and **E** contribute to the force. The particle must be moving to interact with the magnetic field.

- **Motion of charged particle in a magnetic field:** A magnetic field displaces a charged particle perpendicular to the direction of B; the field alters the direction of the particle, but the speed stays the same.
- For a velocity vector perpendicular to B, the particle will be deflected into a circular path whose plane is perpendicular to B.

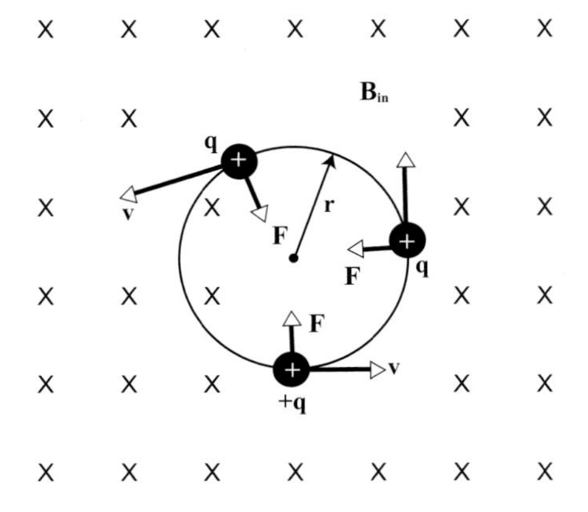

- The **radius of the circular path** is given by r:

$$r = \frac{mv}{qB}$$

- **m** is the **mass** of the particle.
- **v** is the **velocity.**
- **q** is the **charge.**
- **B** is the **magnetic field strength.**

- If the particle is also moving along the direction of B, this path will be **helical,** as shown in figure below:

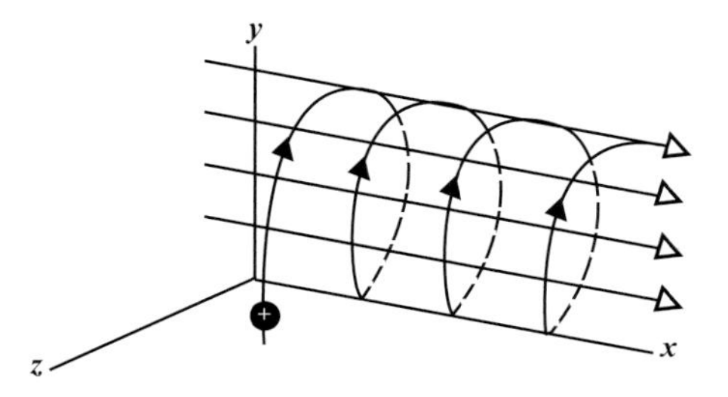

- **Cyclotron Frequency:** The angular frequency of the rotating charged particle is the cyclotron frequency, **ω**, given by the following expression:

$$\omega = \frac{qB}{m}$$

Sources of Magnetic Fields

- A current passing through a conductor generates a magnetic field, B.

Biot-Savart Law

- Current generates a magnetic field.
- Given the current I and the conductor segment of length dl, the induced magnetic field contribution, dB, is described by the following:

$$dB = \frac{\mu_0}{4\pi} \frac{I\, dl \times r}{r^3}$$

- The total magnetic field is given by:

$$B = \frac{\mu_0}{4\pi} I \int \frac{dl \times r}{r^3}$$

- The integration is for the entire conductor.
- **Key Properties of B**
- The **magnetic field strength** varies as the inverse square of the distance from the conducting element.

NOTES:
This mathematical behavior is also observed for the electric field intensity generated by a point charge.

- **Special Case:** Infinitely long straight conducting wire:

$$B(a) = \frac{\mu_0}{2\pi} \frac{I}{a}$$

  - **a** is the **distance** from the wire.
  - **I** is the **current** passing through the conductor.
  - The **field strength, B,** is *inversely proportional* to **a.**

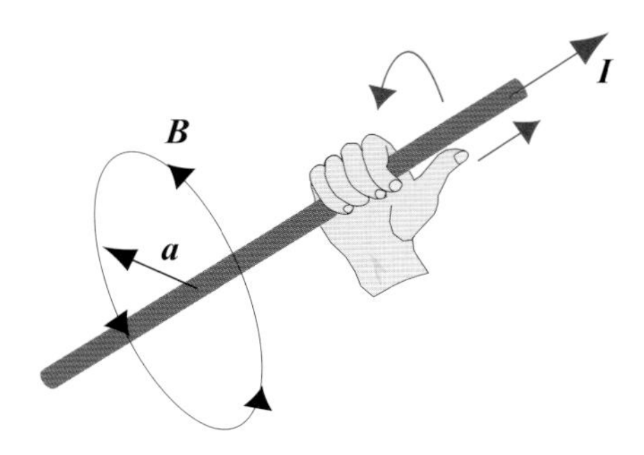

Ampere's Law

- If we focus on a circular path around wire, we find that the total of the magnetic flux, B dot dS, must be consistent with the current passing through the wire, I:

$$\oint \mathbf{B} \cdot d\mathbf{S} = \mu_0 \mathbf{I}$$

- **Magnetic Flux: φ_m**
- The magnetic flux, φ_m, associated with an area, dA, of an arbitrary surface is given by the following equation:

$$\varphi_m = \int \mathbf{B} \cdot d\mathbf{A}$$

- dA is a vector perpendicular to the area dA.
- **Special Case:** Planar area A and uniform B at angle θ with dA.

$$\varphi_m = B\,A\cos\theta$$

Gauss's Law

- The net magnetic flux through any closed surface is always zero.

  $$\oint \mathbf{B}\cdot d\mathbf{A}=0$$

- Gauss's Law is based on the observation that isolated magnetic poles (called **monopoles**) have not been detected experimentally.

Faraday's Law & Electromagnetic Induction

- Passing a magnet through a current loop induces a current in the loop.

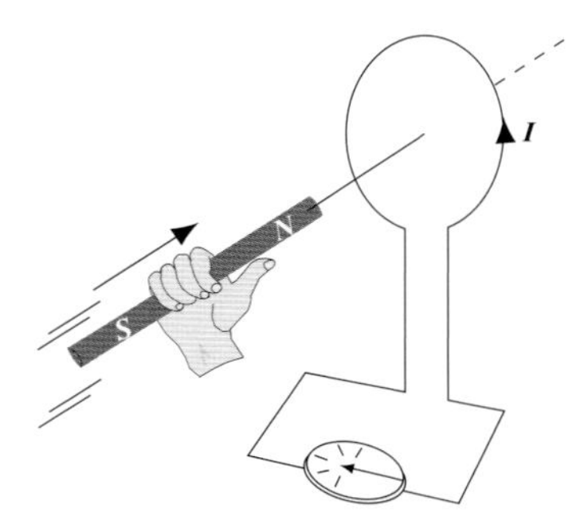

Key Equations

- **Faraday's Law of Induction:** The EMF induced in a circuit is directly proportional to the time rate of change of the magnetic flux, φ_m, passing through the circuit:

  $$\mathrm{EMF}=\oint E\, ds=\frac{-d\varphi_m}{dt}$$

- φ_m is the magnetic flux, given by the following expression:

$$\varphi_m = \int B \cdot dA$$

- The integration covers the area bounded by the circuit, and accounts for variations in the electric field, E.
- **Special Case:** Uniform field B over loop of area A. θ is the angle formed by dA and B:

$$\mathbf{EMF = -d/dt\ (BA\cos\theta)}$$

- **Motional EMF:** Moving a conducting bar of length l through a magnetic field B with a speed v induces an EMF (B is perpendicular to the bar and v):

$$\mathbf{EMF = -\ B\ l\ v}$$

Lenz's Law

- The direction of the induced current and EMF tends to maintain the original flux through the circuit.
- Lenz's Law is a consequence of energy conservation.

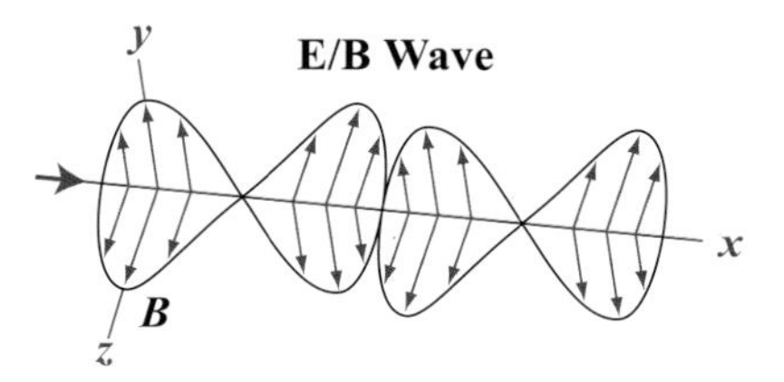

Electromagnetic Waves

Key Equations & Concepts

- Electromagnetic waves are formed by transverse B and E fields.
- The relative field strengths are defined by the following equation:

$$\frac{E}{B} = c$$

- The speed of light, c, is also found to correlate the magnetic constant, μ_0, the permeability of free space, and the electric constant, ε_0, the permittivity of free space:

$$c = \frac{1}{\sqrt{\mu_0 \varepsilon_0}}$$

- In a vacuum, an electromagnetic wave travels at the speed of light, c.
- The wave properties are governed by the following equation:

$$c = f\lambda$$

- **Electromagnetic waves** are found with a wide range of λ.
- **X-rays** have short wavelength, compared with the other extreme of long wavelength radio waves.
- **Visible light** comprises a very small part of the spectrum.

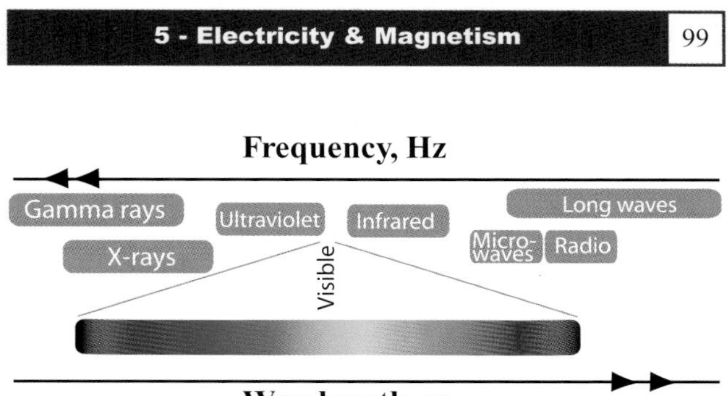

Maxwell's Equations

- Maxwell's Equations summarize the general behavior of electrical and magnetic fields in free space.
- **Gauss's Law for Electrostatics:**

$$\oint \mathbf{E} \cdot d\mathbf{A} = \frac{Q}{\varepsilon_0}$$

Key: Electrostatic charge gives rise to the electric field, **E**

- **Gauss's Law for Magnetism:**

$$\oint \mathbf{B} \times d\mathbf{A} = 0$$

Key: Absence of magnetic monopoles

- **Ampere-Maxwell Law:**

$$\oint \mathbf{B} \cdot d\mathbf{s} = \mu_0 I + \mu_0 \varepsilon_0 \frac{d\varphi_e}{dt}$$

Key: A magnetic field is produced by current and a change in electric flux

■ Faraday's Law:

$$\int \mathrm{E} \cdot \mathbf{dS} = \frac{-\mathbf{d}\boldsymbol{\mu}_{\mathrm{m}}}{\mathbf{dt}}$$

Key: The change in magnetic flux produces an electric field, **E**

6 Light

Basic Properties of Light

Goal: Examine light and its interaction with matter. Light exhibits a duality—having both wave and particle properties.

Key Variables

- **Speed of light** in a vacuum
- **Index of refraction, n:**
- The speed of light decreases as it passes through transparent materials.
- The index of refraction, symbolized n, is the ratio of the speed of light in a vacuum divided by the speed of light in the material:

$$n = \frac{c\,(\text{vacuum})}{c\,(\text{material})}$$

- View **light as a wave:** Focus on wave properties, wavelength and frequency.
- For light as an **electromagnetic wave:**

$$\lambda f = c$$

- Light is characterized by its "color" or **wavelength,** or by its **frequency**, **f**.

- View **light as a particle** in order to understand the energetic properties of light.
- Energy is quantized in packets, called **photons.**
- The energy of a photon of light is proportional to the frequency, f, with the proportionality constant, h:

E (photon) = h f

NOTES:
This is also known as **Planck's Constant.**

Reflection & Refraction of Light

sin & cos waves

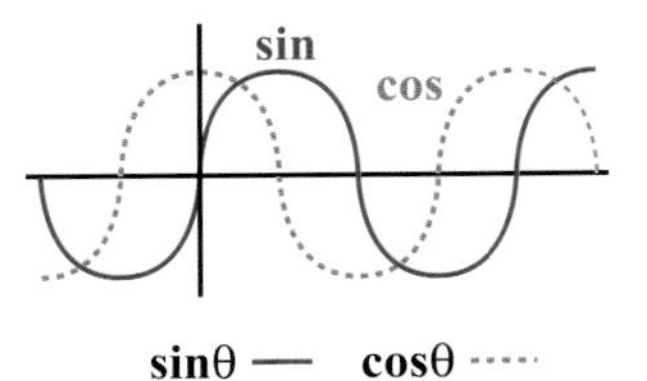

sinθ — **cosθ** ·····

- **Law of Reflection:** For light reflecting from a mirrored surface, the incident and reflected beams must have the same angle with the surface normal:

$\theta_1 = \theta_r$

Reflection & Refraction

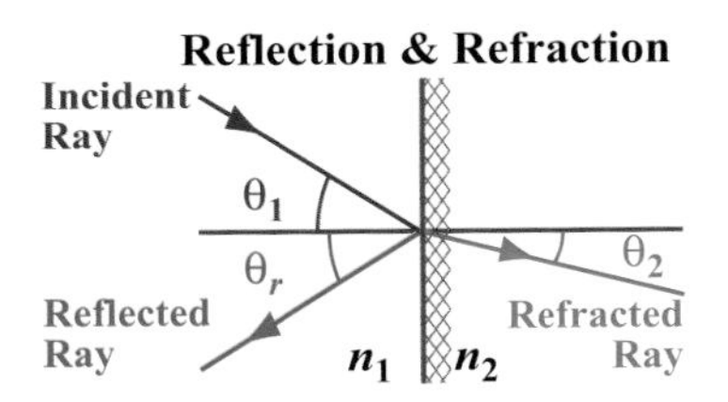

- **Refraction:** Light changes speed as it passes through materials with different indices of refraction.
- This change in speed bends the light ray as it passes from n_1 to n_2.
- The angles of the incident and refracted rays are governed by **Snell's Law:**
  $\mathbf{n_1 \sin \theta_1 = n_2 \sin \theta_2}$
- n_1, n_2: indices of refraction of two materials

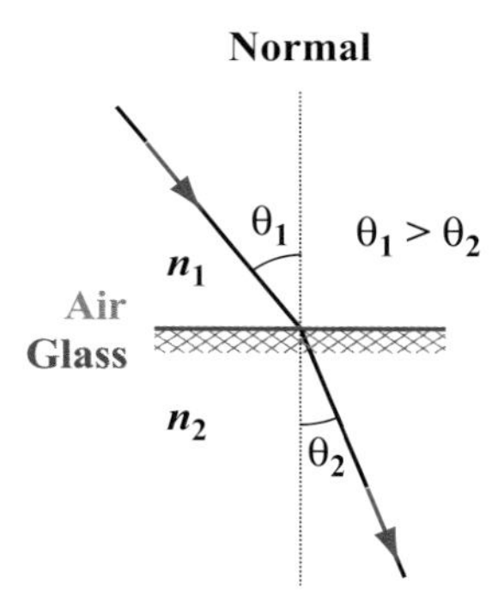

- **Internal Reflectance:** $\sin \theta_c = \frac{n_2}{n_1}$
- Light passing from material of higher n to a lower n may be trapped in the material if the angle of incidence is too large.
- **Polarized Light:** The E field of the electromagnetic wave is not spherically symmetric in polarized light.

> **SAMPLE:**
> Plane (linear) polarized light, circularly polarized light

- One way to generate a polarized wave is by reflecting a beam on a surface at a precise angle, called $\boldsymbol{\theta}_c$.
- The angle depends on the relative indices of refraction and is defined by **Brewster's Law:**

$$\tan \theta_c = \frac{n_2}{n_1}$$

Lenses & Optical Instruments

Goal: Lenses and mirrors generate images of objects.

Key Concepts & Variables

- Lenses and mirrors are characterized by a number of optical parameters.
- The **radius of curvature,** R, defines the shape of the lens or mirror; R is two times the **focal length,** f: **R = 2 f**
- The **optic axis:** line from base of object through center of lens or mirror.

Lens & Mirror Properties		
Parameters	+ sign	- sign
f focal length	converging lens concave mirror	diverging lens convex mirror
s object distance	real object	virtual object
s' image distance	real image	virtual image
h object size	erect	inverted
h' image size	erect	inverted

- **Magnification:** The magnifying power of a lens is given by **M,** the ratio of image size to object size:

$$M=\frac{h'}{h}$$

Laws of Geometric Optics

- **The mirror equation:** The focal length, image distance and object distance are described by the following relationship:

$$\frac{1}{s}+\frac{1}{s'}=\frac{1}{f}$$

- The object and image distances can also be used to determine the magnification:

$$\frac{s}{s'}=-\frac{h}{h'}=M$$

- A **combination of two thin lenses** gives a lens with properties of the two lenses.
- The focal length is given by the following equation:

$$\frac{1}{f}=\frac{1}{f_1}+\frac{1}{f_2}$$

- **General guidelines for ray tracing:**

  - Rays that parallel optic axis pass through "f."
  - Rays pass through center of the lens unchanged.

◆ Image: formed by the convergence of ray tracings.

■ Illustrated **ray tracings**:

Plane Mirror – Law of Reflection

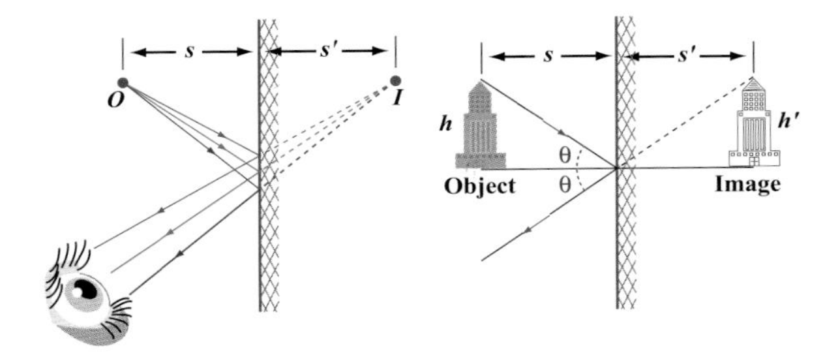

Converging Lens

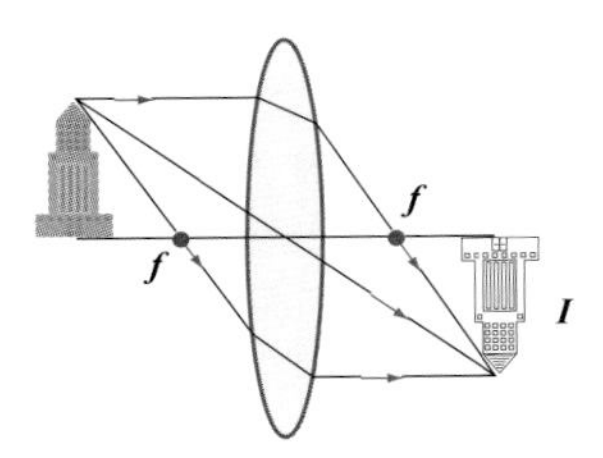

Spherical Concave Mirror

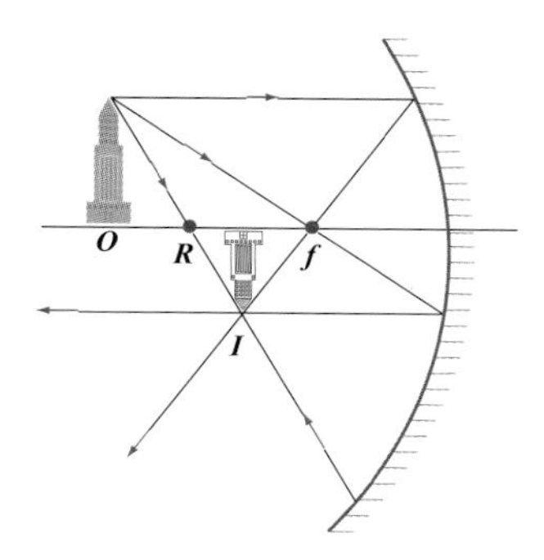

Interference of Light Waves

Goal: Examine constructive and destructive interference of light waves.

Key Variables & Concepts

- **Constructive Interference:** Occurs when wave amplitudes add up to produce a new wave with a larger amplitude than either of the component waves.

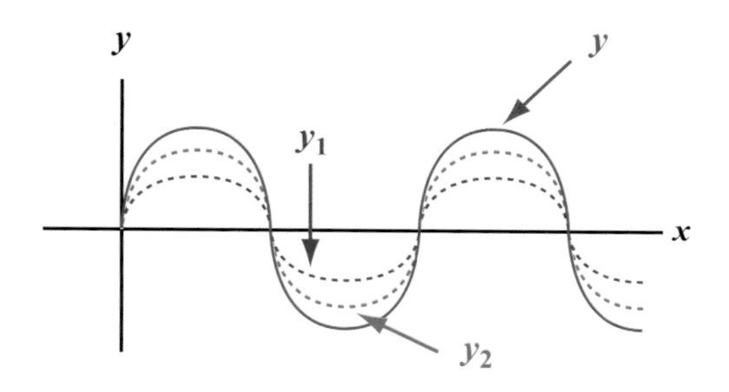

- **Destructive Interference:** Occurs when wave amplitudes add up to produce a new wave with smaller amplitude than either of the component waves. The wave amplitudes cancel out.

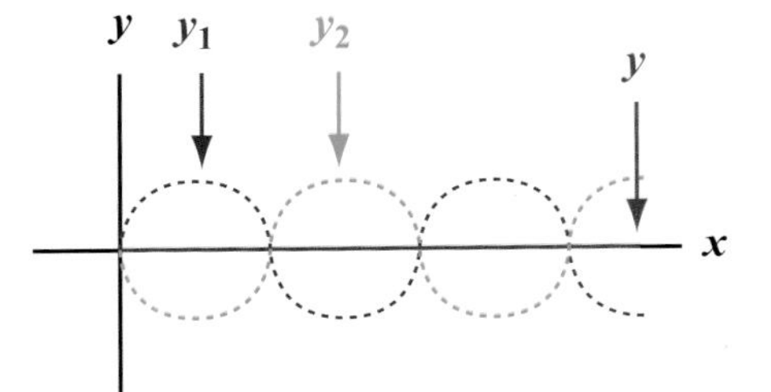

- **Huygens' Principle:** Each portion of wave front acts as a source of new waves.

- **Diffraction of light** from a **grating** with spacing d, produces an interference pattern governed by the following equation:

  $$d \sin \theta = m \lambda, \qquad (m = 0, 1, 2, 3, \ldots)$$

- **Single slit experiment:** For a wave passing through a slit of width a, destructive interference is observed for:

  $$\sin \theta = m \lambda / a, \qquad (m = 0, \pm 1, \pm 2 \ldots)$$

- **X-ray diffraction** from a crystal with atomic spacing d gives constructive interference for:

  $$2 d \sin \theta = m \lambda, \qquad (m = 0, 1, 2, 3, \ldots)$$

7 Modern Physics

Nuclear Physics

- **Atomic Number,** given the symbol Z, is the number of protons in the nucleus of an atom.
- The **atomic number** defines the identity of the element.
- **The Periodic Table of the Elements** [*see* page 110] summarizes all natural and synthetic elements.
- **Mass Number: $A = Z + N$**

  N = number of neutrons in the isotope.
- **Isotopes** are atoms with the same Z, but different A. They are the same element, but have different **atomic masses.**
- Most elements exhibit several natural and synthetic isotopes. For calculations involving mass of materials, the atomic weight is used to denote the mass of the material.

Periodic Table of the Elements

1 H																	2 He
3 Li	4 Be											5 B	6 C	7 N	8 O	9 F	10 Ne
11 Na	12 Mg											13 Al	14 Si	15 P	16 S	17 Cl	18 Ar
19 K	20 Ca	21 Sc	22 Ti	23 V	24 Cr	25 Mn	26 Fe	27 Co	28 Ni	29 Cu	30 Zn	31 Ga	32 Ge	33 As	34 Se	35 Br	36 Kr
37 Rb	38 Sr	39 Y	40 Zr	41 Nb	42 Mo	43 Tc	44 Ru	45 Rh	46 Pd	47 Ag	48 Cd	49 In	50 Sn	51 Sb	52 Te	53 I	54 Xe
55 Cs	56 Ba	57 La*	72 Hf	73 Ta	74 W	75 Re	76 Os	77 Ir	78 Pt	79 Au	80 Hg	81 Tl	82 Pb	83 Bi	84 Po	85 At	86 Rn
87 Fr	88 Ra	89 Ac#	104 Rf	105 Db	106 Sg	107 Bh	108 Hs	109 Mt	110 Ds	111	112	113 Uut	114	115 Uup	116	117	118

58 Ce	59 Pr	60 Nd	61 Pm	62 Sm	63 Eu	64 Gd	65 Tb	66 Dy	67 Ho	68 Er	69 Tm	70 Yb	71 Lu
90 Th	91 Pa	92 U	93 Np	94 Pu	95 Am	96 Cm	97 Bk	98 Cf	99 Es	100 Fm	101 Md	102 No	103 Lr

- The **atomic weight** is a **weighted average** of observed isotopes for each element. [Examples of the isotopes for some of the lighter elements are given in the table below.]

Examples of Isotopes			
Isotope	Atomic Mass	Isotopic Composition	Atomic Weight
H-1	1.0078250321(4)	99.9885(70)	1.00794(7)
H-2	2.0141017780(4)	0.0115(70)	
H-3	3.0160492675(11)		
He-3	3.0160293097(9)	0.000137(3)	4.0026032497(10)
He-4	4.002602(2)	99.999863(3)	
Li-6	6.0151223(5)	7.59(4)	6.941(2)
Li-7	7.0160040(5)	92.41(4)	
Be-9	9.0121821(4)	100	9.012182(3)
B-10	10.0129370(4)	19.9(7)	10.811(7)
B-11	11.0093055 (5)	80.1(7)	
C-12	12.0000000(0)	98.93(8)	12.0107(8)
C-13	13.0033548378(10)	1.07(8)	
C-14	14.003241988(4)		

- **Binding Energy:** The nuclear mass does not equal the sum of proton and neutron masses.
- The mass difference (Δm, kg/mol) is due to the **nuclear binding energy** (ΔE, J):

$$\Delta E = \Delta m\, c^2$$

NOTES:
The units for energy, **J = kg m^2/s^2;** the mass change is in kg.

- **Nuclear reactions** alter the nucleus of an atom.
- Subatomic particles are ejected or absorbed, and new isotopes may be created by the reaction.
- Charge and mass-energy must balance.
- Mass may be converted to energy, following **Einstein's equation:**

$$e = mc^2$$

- **Nuclide symbol:** For element X, the nuclide is symbolized by ${}^{A}_{Z}X$ or by the elemental symbol followed by the mass number, X-A.
- For example, deuterium is one of the isotopes of hydrogen: denoted as H-2 or ${}^{2}_{1}H$.

Nuclear Processes

- **Transmutation:** New elements made from particle collisions.

Common Radioactive Particles	
α	${}^{4}_{2}$**He** nucleus, +2 charge
β	energetic electron, -1 charge
γ	high energy photon or x-ray
Neutron	neutral subatomic particle
Positron	mass of electron, +1 charge

- **Fusion:** Small atoms combine to form larger atoms.
- **Fission:** Large atoms split into smaller atoms and particles.
- **Spontaneous decay** of unstable isotopes is observed for natural radioactive isotopes.
- **Radioactive decay** occurs when an unstable nucleus emits particles as it disintegrates.
- The **rate of radioactive decay** involves **half-life.**
- Emissions are characterized by specific energy and **half-life**:

$$t_{1/2}$$

- Half-life corresponds to the time required for one-half of the nuclei in the sample to decay.

- **Time dependence of decay** is another factor.
- Starting with N0 nuclei, at any time, t, the number of remaining in the sample is given by:

$$N = N_0 e^{\lambda t}$$

- **λ,** the decay constant, is related to $t_{1/2}$:

$$\lambda = \frac{0.693}{t_{1/2}}$$

- **Activity,** A, called the **disintegration rate,** measures the rate of decay of the isotope, dN/dt.
- Because this is a first order process:

$$A = \lambda N$$

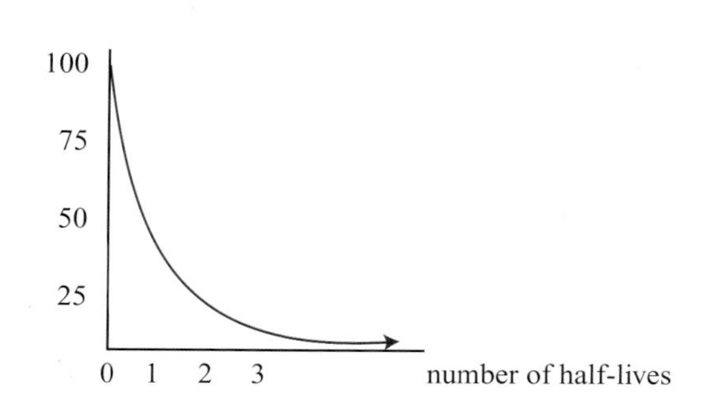

Special Relativity

Special Relativity is a revision of Newton's Laws of Motion for bodies moving at a velocity approaching the speed of light, c.

Special Relativity Postulates

- **Postulate 1:** All laws of physics are the same in every inertial frame of reference.
- **Postulate 2:** The speed of light is constant, regardless of the motion of the source or observer.
- These postulates have ramifications on the **mass, energy** and **time** measured by an observer in a reference frame separate from the body in motion.

Lorentz Transformations: These mathematical expressions describe an event in system, S, as experienced by an observer in another system, S', moving with constant velocity, u, along the x-axis (could also treat motion along *y*-axis or *z*-axis).

$$x' = \frac{(x - ut)}{\sqrt{1 - \left(\frac{u}{c}\right)^2}}$$

$$t' = \frac{\left(t - \frac{ux}{c^2}\right)}{\sqrt{1 - \left(\frac{u}{c}\right)^2}}$$

- **Consequences of Relativity:** For an observer moving at velocity, u, relative to the frame of reference of object or event, length, time and mass exhibit relativistic effects:

  - **Length contraction:**

    $$\text{length (observed)} = \text{length}\sqrt{1 - \left(\frac{u}{c}\right)^2}$$

  - **Time dilation:**

    $$\text{time (observed)} = \frac{\text{time}}{\sqrt{1 - \left(\frac{u}{c}\right)^2}}$$

  - **Mass increases:**

    $$\text{mass (observed)} = \frac{\text{mass}}{\sqrt{1 - \left(\frac{u}{c}\right)^2}}$$

NOTES:
These effects are only observable when the velocity approaches the speed of light; otherwise, the effects are inconsequential.

- **Mass-Energy Conservation**
- The mass and energy of a body change as the speed of a body, u, approaches the speed of light, c.

- **Rest mass = m_0**
  Therefore, the rest energy, $\mathbf{E_0 = m_0 c^2}$
- **Relativistic mass = m**
  Therefore, relativistic energy, $\mathbf{E = m c^2}$

Hint: $m = \dfrac{m_0}{\sqrt{1-(\frac{u}{c})^2}}$

- The difference between the rest energy and the relativistic energy is K, the **kinetic energy:**

$$K = E - E_0 = \frac{m_0 c^2}{\sqrt{(1-(\frac{u}{c})^2)}} - m_0 c^2$$

Quantum Theory

> **NOTES:**
> Under the principles of relativity, the **total energy** is conserved for any process, not the **kinetic energy** or **rest energy.**

- **Light as a Particle:** Light exhibits wave and particle properties, depending on how you measure the behavior.
- The energy of electromagnetic radiation is carried in definite bundles of energy called **photons.**
- The energy of photon is given by:
  $\mathbf{e = h f}$

  - h is Planck's Constant.
  - f is the frequency of the light.

- **Photon Momentum:** Even though the photon is massless, it carries a **quantized momentum,** p, related to λ by the following equation:

$$p=\frac{h}{\lambda}=\frac{hf}{c}$$

  - This is the principle behind the concept of a solar sail for deep space satellites.

- **Wave Character of Particles:** Matter exhibits wave and particle properties, depending on the conditions.
- **The DeBroglie Equation** defines the wavelength, λ, of a body of mass, m, moving with velocity, v:

$$\lambda\,(\text{DeBroglie})=\frac{h}{mv}=\frac{h}{p}$$

  - h is Planck's constant.
  - $m \times v$ is the momentum, p.

- The wave character is observed for subatomic particles, such as electrons in atoms and beams of neutrons.

Experimental Evidence for the Quantum Theory of Matter

- **Photoelectric Effect:** The exposure of a metal surface to a photon of sufficient energy will result in the ejection of electrons from the metal surface.
- The photon energy must exceed a threshold value, called the work function, W.

- If the photon energy is less than W, no electrons are emitted, even for high intensity photon sources.
- The maximum kinetic energy of the ejected electrons, K_{max}, is given by the difference between hf and W: $\mathbf{K_{max} = hf - W}$

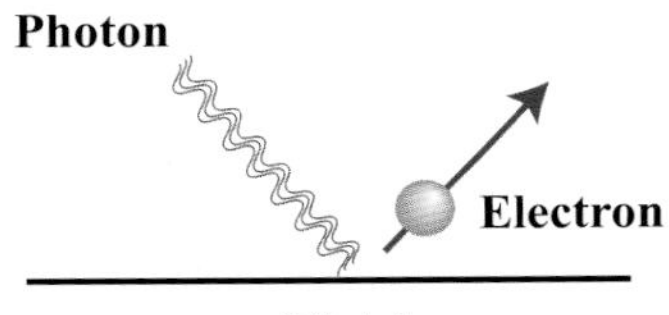

- **Compton Scattering:** The scattering of a photon and an electron will produce a deflected photon of longer wavelength, with the energy gained by the electron.

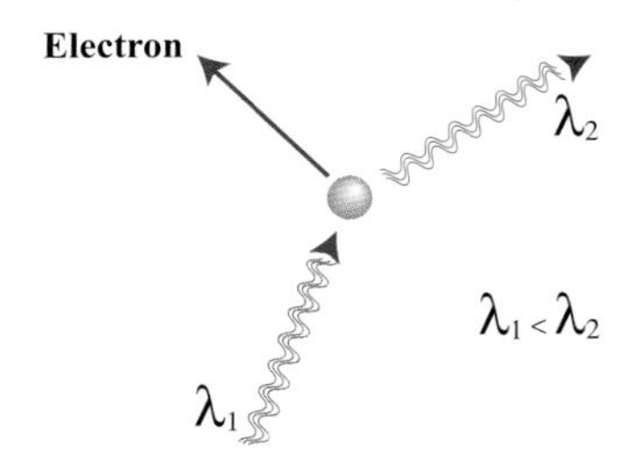

- **Quantum mechanics** provide the methods to examine the wave character and associated energies for particles, atoms, molecules and solids.
- The full description of the quantum mechanical wave character is called the **wavefunction.**
- The **Schrodinger Equation ($H\psi = E\psi$)** describes the energy **(E)** and wavefunction **(ψ)** for any system of interest.

- This is a coupled differential equation: H (the Hamiltonian operator) is the mathematical description of the energy of the system.
- For many electron systems, H depends on the electron density, which is derived from ψ.
- **Hydrogen Atom:** The Bohr model showed that the energy of the electron of hydrogen is quantized, occupying discrete energy states:

$$E(n) = \frac{1}{n^2} E_1$$

  - n is the quantum number.
  - E_1 is the energy of the most stable electronic state.

- This model accurately explained the observed atomic spectra for atomic hydrogen.
- Further applications of quantum mechanics obtained descriptions of the **particle waves,** called **electronic orbitals.**

8 Equations & Answers

Assessing Experimental Data

- **Accuracy:** the agreement between experimental data and a known value.
- **Error:** a measure of **experimental error:**
  - ◆ Error = (experiment value - known value)
  - ◆ Relative error = Error / (known value)
  - ◆ % Error = Relative Error x 100%

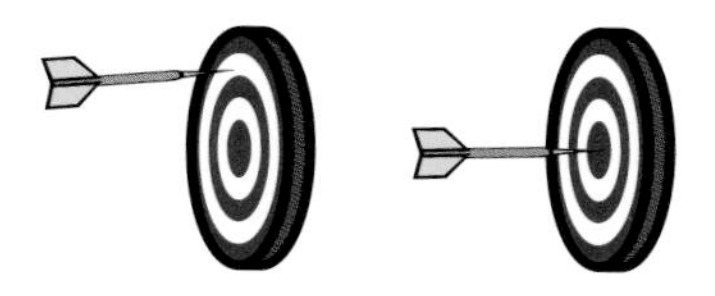

> **SAMPLE:**
> The experimental value is 5.51, the known value is 5.80. What is the % error?
> The error = 5.51- 5.80 = -0.29
> The relative error is -0.29/5.8 = - 0.050
> % error: - 0.05*100% = - 5.0%
> In this case, we undershoot the known value.

- **Precision:** the degree to which a set of experimental values agree with each other.

> ⚠ **PITFALL:** a set of data can be precise, but have a large experimental error.

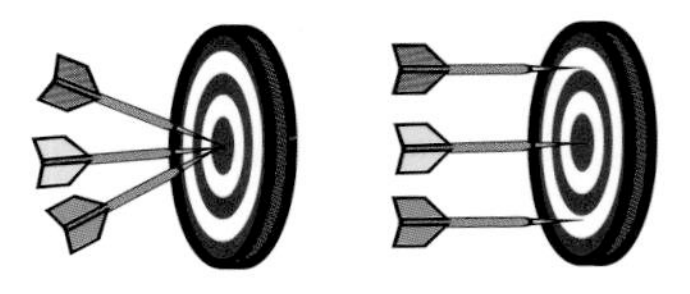

Calculation of Statistical Parameters

- **For a set of numbers, $\{x_1, x_2, x_3 \dots x_j\}$**
  - ◆ The **mean** or average value is the sum of all "x" divided by "j."
  - ◆ The deviation, **Δ**, for each x is "x - mean"; **Δ** can be positive or negative.

SAMPLE:

x	25	28	31	35	43	48
Mean = (25+28+31+35+43+48)/6 = 35						
Δ	-10	-7	-4	0	+8	+13

Note: The sum of **Δ** = 0

Significant Figures (sigfig)

- The number of sigfigs reflects the accuracy of experimental data; calculations must accommodate this uncertainty.
- **For multiplication:** The number of sigfigs in the final answer is limited by the entry with the fewest sigfigs.
- **For addition:** the number of decimal places in the final answer is given by the entry with the fewest decimal places.
- Rules for "rounding sigfigs":
  - ◆ If the last digit is >5, round up.

◆ If the last digit is <5, round down.
◆ If digit = 5; round up, if preceding digit is odd.

SAMPLE:
1.245 + 0.4 = 1.6 (1 decimal place)
1.345 × 2.4 = 3.2 (2 sigfigs)

◆ Record all the certain-digits and one uncertain or estimated digit for a measurement.
◆ The calculator often includes extra digits.
◆ For a multi-step problem, keep 1 or 2 extra digits; then round off the **final** answer.

◆ Rules for the number of sigfigs in the final answer:
- For **addition** or subtraction, use the **least** number of decimal places found in the data.

SAMPLE:
10.102 + 5.03 = 15.13

- For **multiplication** or division, the final answer should have same number of sigfigs as the entry with the **smallest** number of sigfigs.

◆ **Rounding-off data:** Round up if the last **uncertain** digit is 6,7,8,9; round down if it is 0,1,2,3,4. If it is a 5, the **arbitrary** convention is to round up if the last **certain** digit is odd, round down if it is even.

SAMPLE:
5.462 × 4.00 = 21.8

> **SAMPLE:**
> 5.085 rounds off to 5.08
> 0.035 rounds off to 0.04
> 1.453 rounds off to1.45
> 20.248 rounds off to 20.25

Basic Skills

Calculator Survival

- Become familiar with your calculator **before** the exam; make sure you can multiply, divide, add, subtract and use all needed functions.

- Calculators never makes mistakes: They take your input (intended & accidental) and give an answer.

- Look at the answer: Does it make sense?

- Do a quick estimate as a check on your work.

> **SAMPLE:**
> $\frac{4.34 \times 7.68}{1.05 \times 9.8}$ is roughly $\frac{4 \times 8}{1 \times 10}$ or 3.2;
> actual answer: 3.24

Word Problems

- **Read** and evaluate the question **before** you start plugging numbers into the calculator.

- Identify the variables, constants and equations.
- Write out units of the variables and constants.
- Work out the unit before the numerical part.
- You may have extra information, or you may need to obtain constants from your text.

All numerical data has units. In physics, we use **metric** "SI" units. Other units must be converted to SI units in order to be used with standard equations and constants.

⚠ **PITFALL:** If the unit is wrong, the answer is wrong!

Appendix

Fundamental Physical Constants		
Mass of Electron	m_e	$9.11 x10^{-31}$ kg
Mass of Proton	m_p	$1.67 x10^{-27}$ kg
Avogadro Constant	N_A	$6.022 x10^{23}$ mol^{-1}
Elementary Charge	e	$1.602 x10^{-19}$ C
Faraday Constant	$\mathcal{F}$	96,485 C/mol
Speed of Light	c	$3 x10^8$ ms^{-1}
Molar Gas Constant	R	8.314 J mol^{-1} K^{-1}
Boltzmann Constant	k	$1.38 x10^{-23}$ JK^{-1}
Gravitation Constant	G	$6.67 x10^{-11}$ m^3 $kg^{-1}s^{-2}$
Permeability of Space	μ_0	4π x 10^{-7} N/A^2
Permittivity of Space	ε_0	8.85 x 10^{-12} F/m

Conversion Factors & Alternative Units		
Angle	° degree	180° = π rad
Energy	Erg	CGS unit (g cm^2/s^2) 1 erg = 10^{-7}J
Energy	Electron Volt	1 eV = 1.602 $x10^{-19}$J

Conversion Factors & Alternative Units *(continued)*		
Force	Dyne	CGS unit ($g\ cm/s^2 = erg/cm$) 1 dyne = 10^{-5} N
Volume	Liter	1L = $1dm^3$
Pressure	Bar	1 Bar = 10^5 Pa
Length	Angstrom	1 Å = $1x10^{-10}$m

Units of Basic Variables	
time: second s	position: meter m
mass: kilogram kg	volume: m^3
density: kg/m^3	temperature, K
velocity m/s	acceleration: m/s^2
energy: Joule $J = kg\ m^2/s^2$	force: Newton $N = kg\ m/s^2$

Greek Alphabet		
	Capital	Lowercase
Alpha	A	α
Beta	B	β
Gamma	Γ	γ
Delta	Δ	δ
Epsilon	E	ε
Zeta	Z	ζ
Eta	H	η
Theta	Θ	θ

Greek Alphabet *(continued)*		
	Capital	Lowercase
Iota	I	ι
Kappa	K	κ
Lambda	Λ	λ
Mu	M	μ
Nu	N	ν
Xi	Ξ	ξ
Omicron	O	ο
Pi	Π	π
Rho	P	ρ
Sigma	Σ	σ
Tau	T	τ
Upsilon	ϒ	υ
Phi	Φ	φ
Chi	X	χ
Psi	Ψ	ψ
Omega	Ω	ω

Units for Physical Quantities		
Basic Physical Quantity	Symbol	Unit
Length	l,x	Meter - m
Mass	m,M	Kilogram - kg
Temperature	T	Kelvin - K
Time	t	Second - s
Electric Current	I	Ampere - A(C/s)

Derived Quantities	Symbol	Unit
Acceleration	a	m/s^2
Angle	θ, φ	radian
Angular Acceleration	α	$radian/s^2$
Angular Momentum	L	kg m 2/s
Angular Velocity	ω	radian/sec
Capacitance	C	Farad, F (C/V)
Charge	Q, q, e	Coulomb, C(As)
Density	ρ	g/cm^3 or kg/m^3
Displacement	s,d,h	Meter, m
Electric Field	E	V/m
Electric Flux	ϕ_e	V m
Electromotive Force (EMF)	$\mathcal{E}$	Volts, V
Energy	E, U, K	Joule, J(kg m^2 s^{-2})
Entropy	S	J/K
Force	F	Newton, (kg m/s^2 = J/m)
Frequency	f	Hertz, Hz(cycle/s)
Heat Energy	Q	Joule, J

Derived Quantities	**Symbol**	**Unit** *(continued)*
Magnetic Field	B	Tesla, T (Wb/m^2)
Magnetic Flux	Φ_m	Weber, Wb ($kg\ m^2/A\ s^2$)
Momentum	p	m x v = kg m/s
Potential	V	Voltage, V(J/C)
Power	P	Watt, W(J/s)
Pressure	P	Pascal, Pa (N/m^2)
Resistance	R	Ohm, Ω(V/A)
Torque	τ	N m
Velocity	v	m/s
Volume	V	m^3
Wavelength	λ	Meter, m
Work	W	Joule, J(N m)

Unit Prefix - denotes power of "10" of SI units		
Prefix	**Symbol**	**Factor**
peta	P	10^{15}
tera	T	10^{12}
giga	G	10^{9}
mega	M	10^{6}
kilo	k	10^{3}
hecto	h	10^{2}
deka	da	10^{1}
deci	d	10^{-1}
centi	c	10^{-2}
milli	m	10^{-3}
micro	μ	10^{-6}
nano	n	10^{-9}
pico	p	10^{-12}
femto	f	10^{-15}
atto	a	10^{-18}

The prefix symbol is printed before the SI unit symbol. A prefix only has meaning when used with an SI unit.

SAMPLE:

1.10 mg = 1.10 x 10^{-3} g

In performing calculations, data with a modified SI unit must be replaced with the numerical equivalent, or be altered to be consistent with other units and constants.

The mass of an object in 3.4 ng. In the calculations using "mass," 3.4 ng is replaced by 3.4 x 10^{-9} g.